THE OXFORD SHAKESPEARE

General Editor · Stanley Wells

THE OXFORD SHAKESPEARE

The Tragedy of Macbeth

EDITED BY NICHOLAS BROOKE

CLARENDON PRESS · OXFORD

1990

Oxford University Press, Walton Street, Oxford OX2 6DP

Oxford New York Toronto
Delhi Bombay Calcutta Madras Karachi
Petaling Jaya Singapore Hong Kong Tokyo
Nairobi Dar es Salaam Cape Town
Melbourne Auckland

and associated companies in
Berlin Ibadan

Oxford is a trade mark of Oxford University Press

Published in the United States
by Oxford University Press, New York

British Library Cataloguing in Publication Data

Shakespeare, William 1564–1616
Macbeth.—(The Oxford Shakespeare)
I. Title   II. Brooke, Nicholas
822.3'3
ISBN 0–19–812901–7
ISBN 0–19–281441–9 (Pbk)

Library of Congress Cataloging in Publication Data

Shakespeare, William, 1564–1616.
Macbeth / edited by Nicholas Brooke.
—(The Oxford Shakespeare)
I. Brooke, Nicholas.   II. Title.   III. Series: Shakespeare,
William, 1564–1616. Works. 1982.
PR2823.A2B76   1990
822.3'3—dc20   89–23858
ISBN 0–19–812901–7
ISBN 0–19–281441–9 (Pbk)

Set by H Charlesworth & Co Ltd
Huddersfield
Printed in Great Britain by
The Alden Press, Oxford

*To Angus Wilson and Tony Garrett
for their enduring friendship
and to Angus for the greatest
English novels of my time*

PREFACE

THIS edition of *Macbeth* has taken over ten years to prepare, an inordinately long time, due to a variety of interruptions as well as to my own dilatoriness. It would have been longer still without the benefit of Kenneth Muir's Arden edition, originally published as far back as 1951, but finally revised with a new Introduction as recently as 1984; I am deeply indebted to Professor Muir, both for that volume, and for his kindness, encouragement, and friendship over many years. An editor's first debt is always to his predecessors; the essential reference work for *Macbeth* is the New Variorum of 1901, faithfully chronicling the variants of all scholarly predecessors from the beginning of the eighteenth to the end of the nineteenth centuries. My successors will hope for an extension through the very active scholarship of another, nearly exhausted, century, and they could be much helped by a similar digest of the numerous surviving prompt-books and editions derived from prompt-books. Drama is an ephemeral art-form, and the changes made in text and structure are constantly revealing of the theatrical potentialities of even so great a text as this: *Macbeth* belongs to the theatre, and I have learnt from far more productions than I can mention in the Introduction.

Fifty years ago *Macbeth* was put to sleep in my mind by the all-too-common unhappiness of studying it for School Certificate; it seemed as impossible to wake as the Sleeping Beauty, despite my profession. Twenty-five years later, Professor Wolfgang Clemen had the generous idea of using a grant from the Volkswagen Institute to invite 'younger Shakespearians' to Munich for a few weeks to visit his Department and its Shakespeare Library, and to be shown the City and its environs. In the celebrated church of St. John Nepomunk I experienced a conversion, not of a religious kind, but to a perception of baroque art. What was most strange was that it was *Macbeth* which came so powerfully into my mind: for better or for worse, my debt to that occasion will be obvious in this volume.

The general editor of this series, Professor Stanley Wells, has been consistently attentive, patient, and kind; his learning has saved me from many follies, and his generosity has allowed me to persist in some minor misdemeanours. Correspondence with Dr Gary Taylor was most helpful with the idiosyncrasies of the original compositors, and Mrs Christine Buckley was helpful with editorial policy. More recently I have been particularly grateful to Frances Whistler of the Arts and Reference Division of the Oxford University Press for the warm support that transforms cliffs into bowling greens; and to Jane Robson whose copy-editing dotted my erratically Weïrd Sisters with remarkable care and consideration. Dr R. V. Holdsworth allowed me to see his unpublished work on Middleton's collaborations with Shakespeare, and he and his wife were lavish in hospitality while he conducted me through a host of notes for his study of *Macbeth*. The Music Adviser to this series, Dr F. W. Sternfeld, gave valuable advice on the original music for the play and put me in touch with Dr Christopher Field who proved a most lucid guide through the maze of misinformation which once surrounded the later music used in the eighteenth and nineteenth centuries. The sources of illustrations are acknowledged below, but I must mention here the friendly and helpful letters of the great theatrical photographer, Angus McBean.

To all these, and to many others whose names appear in later pages, I have much reason to be grateful; and I must acknowledge too a further host of critics, historians, novelists, parodists, and friends whose ideas have long since become part of my own mind so that I cannot know from where or from whom they came. I can, however, recognize a very long-standing debt to my historian brother, Professor Christopher Brooke, who added to it by reading and commenting on parts of the Introduction as well as answering emergency calls for help. Still more personal is my debt to my wife, Julia Lacey Brooke, for her sympathy which shares in so much of my enthusiasm for poetry and theatre, and constantly turns horrid imaginings to present laughter.

NICHOLAS BROOKE

Norwich, Thessaloniki, Harwood-in-Teesdale, and Spitalfields,

August 1989

CONTENTS

List of Illustrations   xi

Introduction   1
1. *Illusion*   1
2. *Language*   7
3. *Baroque Drama*   22
4. *Staging*   34
5. *Text*   49
   (i) *Lineation*   50
   (ii) *Act 3, Scene 5–Act 4, Scene 1*   51
   (iii) *The Brevity of Macbeth*   55
6. *Middleton and the Revision of Macbeth*   57
7. *Dates*   59
   (i) *The Original Play*   59
   (ii) *The Revised Version*   64
8. *Sources*   66
   (i) *Chronicles*   67
   (ii) *Stuart Politics*   71
   (iii) *Classical Ladies*   76
   (iv) *Weïrd Sisters*   78
   (v) *Devil Porters*   79

Editorial Procedures   83
Abbreviations and References   87

MACBETH   1

APPENDIX A
Lineation   213

APPENDIX B
Musical Additions   225

APPENDIX C
Macbeth at the Globe, 1611   234

Index   237

LIST OF ILLUSTRATIONS

1. *The Conversion of Saul*, by Caravaggio   25
   (*S. Maria del Popolo, Rome*)

2. *The Martyrdom of St Matthew*, by Caravaggio   27
   (*S. Luigi Francesi, Rome*)

3. *The Cornaro Chapel*, with *The Ecstasy of St Teresa* by
   Bernini   29
   (*Cornaro Chapel, S. Maria della Vittoria, Rome, by
   permission of the Conway Library, Courtauld Institute of Art*)

4. *The Ecstasy of St Teresa*, by Bernini: central group   30

5. *The Ecstasy of St Teresa*, by Bernini: Teresa's face   30
   (*Figures 1–2, 4–5: photos by permission of the Mansell Collection*)

6. *The Apotheosis of James I*, by Rubens   33
   (*Whitehall Banqueting House, London: photo Crown copyright,
   by permission of the Controller of Her Majesty's Stationery Office*)

7. Betterton as Macbeth with Witches, show of kings, and
   Banquo's Ghost: frontispiece to Rowe's edition, 1709   40

8. Garrick and Mrs Pritchard in Act 2, Scene 2   42
   (*By permission of The Garrick Club, London, and E. T. Archive*)

9. Sarah Siddons in Act 5, Scene 1, Covent Garden, 1812   42

10. Design for Hecate and Spirits by George Scharf, Covent
    Garden, 1837   45

11. Charles and Mrs Kean in Act 2, Scene 2, Princess's
    Theatre, London, 1853   45

12. Design for Chorus of Witches by T. Grieve, Princess's
    Theatre, London, 1853   46
    (*Figures 7–12 by permission of the Shakespeare Centre Library,
    Stratford-upon-Avon*)

13. Laurence Olivier and Vivien Leigh at Stratford-upon-
    Avon, 1955   48
    (*Photo: Angus McBean, by his permission*)

14. Banquo (John Woodvine) and Macbeth (Ian McKellen)
    with the Witches in Act 1, Scene 3, directed by Trevor
    Nunn, The Other Place, Stratford-upon-Avon, 1976   48
    (*Photo: Joe Cocks, by permission of the Joe Cocks Studio,
    Stratford-upon-Avon*)

List of Illustrations

15, 16. Robert Johnson's setting of 'Come away, Hecate' in
Act 3, Scene 5, from the Bull Manuscript, MU. MS 782   230–1

    *(Figures 15–16 by permission of the Syndics of the
Fitzwilliam Museum, Cambridge)*

INTRODUCTION

1. *Illusion*

Macbeth was first produced at a time of radical theatrical change in England. It seems to have been written during 1606 (pp. 59–64) and to have been presented at the Globe Theatre fairly late in that year, and so to have been conceived for performance in daylight, in a constantly light space which could not be physically transformed into darkness. Two years later, in 1608–9, Shakespeare's company, the King's Men, took over the Blackfriars Theatre which had been adapted from the hall of the medieval friary and was therefore basically a dark space into which artificial light had to be introduced—which has been the normal state of all European theatres ever since. From then until the London theatres were finally closed in the 1650s after repeated injunctions from Cromwell's government the repertory had to be adapted for performance in both theatres, in both conditions. Shakespeare's last plays, from *The Winter's Tale* to *The Tempest* (including his collaborations with Fletcher) show remarkable ingenuity in devising spectacular effects which could take advantage of the dark theatre and of the experience of the company in participating in Court masques, while still being performable at the Globe. Most of Shakespeare's earlier plays could no doubt have been easily adapted for revival in the new situation since the basic configuration of the stage seems to have been much the same, but *Macbeth* was a special case: about two-thirds of this play written for the daylight theatre is set in darkness.

All theatre depends, in one way or another, on illusion, but *Macbeth* is exceptional in affirming continuously a direct contradiction of the natural conditions: the transformation of daylight into darkness is a *tour de force* which establishes illusion as, not merely a utility, but a central preoccupation of the play, dramatically announced by an opening unique in Shakespeare's plays, the use of the non-naturalistic prologue by the Weïrd Sisters in 1.1. There follows a carefully controlled range of forms of dramatic illusion which needs to be enumerated, not only because it is so frequently mutilated by the naturalistic tradition

1

of modern theatre, but also because it clarifies the study of illusion as a structural foundation of the play.

1. *Darkness in daylight* is established symbolically by torches and candles whose effect depended on the power of theatrical convention to which a modern audience cannot respond so directly as a Jacobean one, but that is greatly extended linguistically by direct statements, allusions, or indirect suggestion of verbal imagery. The sequence of dark scenes is initiated in 1.5 by Lady Macbeth's invocation to the powers of darkness (39–53), and by her later reference to 'This night's great business' (67); it is sustained through all major scenes until the end of 4.1. The Folio text calls for *Hautboys and Torches* at the beginning of 1.6, but that is probably a book-keeper's anticipation of props needed to open 1.7, where they stress the arrival of darkness alongside a dumb-show of preparations for the evening feast; 1.6 opens in dialogue that reverses the illusion of darkness, Duncan and Banquo exchanging descriptions of the castle's pleasant seat, air, jutty, frieze, the martlet's procreant cradle, etc. This is regularly quoted as an example of Shakespeare's use of words to set a scene, but in truth it is not typical; it is quite exceptional in its invitation to detailed visualization; what we must visualize is not there, of course, but the implied daylight literally is. The illusion of darkness can be withdrawn at will (and then resumed), but with the significant irony that Duncan and Banquo misread the signs: there is nothing gentle or procreant here.

2. *The Weïrd Sisters* are visible to us, and to Macbeth, and to the less questionable sight of Banquo (a touchstone of common sense, like Horatio in *Hamlet*, even if less solid). The Sisters cannot be reduced to projections of Macbeth's mind, they are not mere delusions; though just what Macbeth and Banquo *see* is very questionable. Banquo describes them while they are on stage:

> What are these,
> So withered, and so wild in their attire,
> That look not like th'inhabitants o'th' earth
> And yet are on't?   (1.3.39–42)

No wonder presentations in the theatre vary so much, since they cannot be made to 'look like' this; and if they could, Banquo's words would be redundant. Word, here, is against sight: we are bound to see that they are not what Banquo says; but it is more likely that his description influences our perception than that we conclude that his sight is different from ours. The ambiguity extends to their nature: there is still argument as to whether they are supernatural, or merely village witches, strange old women, as Banquo's later words suggest:

> You should be women,
> And yet your beards forbid me to interpret
> That you are so.   (45–7)

They call themselves the Weïrd Sisters, and Banquo and Macbeth refer to them as such; the only time the word 'witch' is heard in the theatre is in l. 6 of this scene, when the First Witch quotes the words of the sailor's wife as the supreme insult for which her husband must be tortured. 'Weird' did not come to its loose modern usage before the early nineteenth century; it meant Destiny or Fate, and foreknowledge is clearly the Sisters' main function. But the nature of their powers is still ambiguous: they are actively malicious to the master o'th' *Tiger*, but have not the power to destroy him:

> Though his bark cannot be lost,
> Yet it shall be tempest-tossed.   (24–5)

They can appear to Macbeth at will (theirs or his), but confine their interference to prediction. All these powers were, of course, attributed to village witches, but the Weïrd Sisters are more decisively supernatural; confusion has largely arisen because the Folio text refers to them in stage directions and speech prefixes as 'witches'. Their ambiguity, of nature and of power, is fundamental to the ambiguities of experience and knowledge which the play develops.

The conflict of words and appearance is repeated at their exit: we must 'see' them go, but we cannot this time (when they are no longer there) verify the description:

BANQUO
> The earth hath bubbles, as the water has,
> And these are of them; whither are they vanished?

MACBETH
> Into the air; and what seemed corporal melted
> As breath into the wind.   (79–82)

3

Whether they actually go into smoke, down a trap, or flying, it cannot 'look like' this; sight ceases to be rationally reliable:

> BANQUO
> Were such things here as we do speak about?
> Or have we eaten on the insane root
> That takes the reason prisoner? (83–5)

3. *The dagger* is an opposite case: the Weïrd Sisters are attested by sight (ours and Banquo's, besides Macbeth's) but are indefinite in form; the dagger is entirely specific in form though not literally seen by anyone—even Macbeth knows it is not there:

> Art thou not, fatal vision, sensible
> To feeling as to sight? Or art thou but
> A dagger of the mind, a false creation
> Proceeding from the heat-oppressèd brain? (2.1.37–40)

This kind of optical illusion is well known, especially in feverish conditions—the brain registers as sight what is not directly stimulated by optic nerve. Macbeth proceeds to confuse perception further by drawing his actual dagger and then seeing the illusory one as still more vivid, with 'gouts of blood, | Which was not so before' (47–8)—which is how the actual one will be in the next scene.

Words play a great part here, but not words alone: the invisible dagger is necessarily created also by his body, gesture, and above all by his eyes, which focus on a point in space whose emptiness becomes, in a sense, visible to the audience.

4. *Banquo's ghost* is different again: it is seen by Macbeth, it was seen by Simon Forman at the Globe in 1610–11, and it has been seen by audiences in most productions since. Thus far it contrasts with the dagger, but it is also in a different case from the Weïrd Sisters because it is seen by no one else on stage:

> LADY MACBETH   When all's done,
> You look but on a stool. (3.4.67–8)

This differs from the dagger because the emptiness here is not of our perceiving, and from the Sisters because here the 'reliable witnesses' contradict our sight. Scepticism, therefore, becomes as questionable as credulity. The whole effect is aborted if, as so often nowadays, no physical ghost appears on stage.

5. *The apparitions* in 4.1 are a climax to this sequence of stage illusion tricks, though the formal elaboration is not technically the most surprising or exciting (it neither requires nor gets the conjuring-trick surprise of the others). It can use elaborate machinery, but it can equally be done with simple effects—cauldron, smoke, a trap, or even less. It is the Weïrd Sisters' fullest scene, and it is their last; but however superbly nasty their incantation and however spectacular what follows, it does not aim at mystification, and the recapitulation of their disappearance in 1.3—Lennox seeing nothing of their departure (4.1.151–2)—does not this time conflict with our sight since he was off-stage at the time.

From this point on there is a radical change in the presentation of illusion: rational sight progressively displaces potential deception.

6. *Lady Macbeth's sleep-walking* in 5.1 is essentially about delusion, but caused by psychological disturbance not by supernatural agency; our recognition of a natural phenomenon is endorsed by Doctor and Nurse, who also recognize a connection with guilt dreams in its jumble of displaced memories. The mysterious is being progressively dissipated and is finally eliminated in

7. *Birnam Wood*, whose moving is an exercise in camouflage (5.4.4–7) of a kind which is still a commonplace of infantry tactics. Illusion is being reduced to rational explanation, and the cheating account of Macduff's birth (5.7.45–6) marks the end of this process. There are, in Macbeth's words, 'no more sights' (4.1.170); the audience is given a full explanation of Birnam Wood before the event, and will scarcely be surprised by the revelation of Macduff's birth.

The Tempest follows a remarkably similar pattern through a varied range of stage illusions to their formal climax in the masque of goddesses (4.1.39–142), and thereafter a progressive withdrawal until the final 'magic'—Ferdinand and Miranda discovered playing chess (5.1.173.1)—is magical only to that part of the stage audience which believed them dead; to us it requires only the pulling of a curtain. At the end, Prospero's epilogue has the actor asking for applause to release him finally from his role. But *The*

Tempest opens with an exceptional display of realism, the presentation of a shipwreck on stage, which is then immediately revealed as the dramatic illusion which, of course, it has to be:

MIRANDA
If by your art, my dearest father, you have
Put the wild waters in this roar, allay them. (1.2.1–2)

That calls immediate attention to the nature of dramatic illusion, and establishes it as mediator between Magus controlling his spirits, and naturalistic rationalism. In *Macbeth* it is the opening by the Weïrd Sisters which proposes a relation between supernatural and natural phenomena. No amount of quotation from King James's early and credulous *Demonology*[1] will transfer the Sisters from a category of belief into one of verifiable knowledge. The Weïrd Sisters are, like Ariel and Caliban, essentially creatures of drama, not merely naturalistic representations of old women.

8. *Macbeth* does not begin with an illusion of realism, but it does end with one: '*Enter Macduff, with Macbeth's head*' (5.7.83.1)—presumably stuck on the end of a pike (see ll. 84–5). That direction proposes a *trompe-l'œil* head, an art like that attributed to Giulio Romano at the end of *The Winter's Tale*, achieved here, no doubt, by a life-mask of Burbage. That final effect is peculiar, for Malcolm, always an equivocal figure, capitalizes briskly on the decapitation 'Of this dead butcher and his fiend-like queen' (l. 99). When last seen sleep-walking, Lady Macbeth was anything but fiend-like, and the only visible butcher here is not Macbeth but the 'heroic' Macduff with the grotesque head he offers to Malcolm's 'Christian' triumph.

9. The eight distinct forms of dramatic illusion discussed so far are all dependent on staging: darkness in light, the Weïrd Sisters, the dagger, Banquo's ghost, the apparitions, Lady Macbeth's sleep-walking, Birnam Wood, Macbeth's head. The ninth, recurrent throughout the play, is the purely verbal creation of a highly visual but unseen world of babes and cherubim, rooky wood, murdering ministers, and horses eating each other; the unusual stress on sensory actuality leaves audiences with an undefined sense of having seen, smelt, touched far more than we have, though, as with Macbeth's dagger, we know there's no such thing.

[1] Edinburgh, 1597; London, 1603.

2. *Language*

The most elaborate instance of spectacular linguistic effect is Macbeth's soliloquy at the opening of 1.7, which is striking not only for the achievement of its climax, but also for the process by which that is arrived at, and the rapid transpositions of language involved. It opens with a notably plain vocabulary, but a syntax so contorted as to amount to word-juggling:

> If it were done when 'tis done, then 'twere well
> It were done quickly;   (1–2)

The tongue-twisting implies mental conflict, with a growing desire to suppress the knowledge of guilt:

> if th'assassination
> Could trammel up the consequence and catch
> With his surcease, success,   (2–4)

The pattern of conditional clauses is extended, but the vocabulary is now mixed with a number of polysyllabic words emphasizing the evasion of thought: 'assassination' and 'consequence' act as euphemisms for murder and guilt which emerge in the word-play that translates 'surcease' into 'success', and so leads to the conditional

> that but this blow
> Might be the be-all and the end-all—here,   (4–5)

The apparent resolution is very similar to ' 'twere well | It were done quickly', and equally superficial; the sentence is apparently complete, but an attempt is made to strengthen it by syntactic doubling of the last word which functions both as end of that clause and beginning of the next (hence the editorial problem of punctuation):

> —here,
> But here, upon this bank and shoal of time,
> We'd jump the life to come.   (5–7)

The monosyllables return to close, finally, the long contorted sentence. But the simplicity is deceptive, for the strong rhetorical gesture allows another suppressed level of thought to emerge, religious fear. It is immediately withdrawn, at the expense of reopening the argument:

> But in these cases
> We still have judgement here,   (7–8)

A third 'here' takes us back to the false hope of resolution, and refutes it, but 'judgement' does not immediately signify the law, or rather is not allowed to, for the obscure words that follow deal apparently only with the inevitability of revenge:

> that we but teach
> Bloody instructions, which being taught, return
> To plague th'inventor.   (8–10)

'Bloody' and 'plague' betray the evasions, and the language expands to reveal the power of what has been evaded:

> This even-handed justice
> Commends th'ingredience of our poisoned chalice
> To our own lips.   (10–12)

The emblematic justice leads to 'ingredience' and 'chalice', both words with strong ecclesiastical associations. Justice and religion have both emerged again, in a language more expansive and elevated than any before. It is, once again, returned to a manageable level, of socio-moral orthodoxy, the ties of kinsman, subject, and host; but what follows after is quite extraordinary:

> Besides, this Duncan
> Hath borne his faculties so meek, hath been
> So clear in his great office, that his virtues
> Will plead like angels, trumpet-tongued against
> The deep damnation of his taking-off;
> And pity, like a naked new-born babe,
> Striding the blast, or Heaven's cherubim, horsed
> Upon the sightless couriers of the air,
> Shall blow the horrid deed in every eye
> That tears shall drown the wind.   (16–25)

It is the process, here, which is all-important: the process of amplification from Duncan's virtues to 'the sightless couriers of the air' is startling in its strangeness. Cleanth Brooks, in a celebrated essay on the naked new-born babe,[1] noted the strangeness but after enumerating the other references to babes in the

---

[1] 'The Naked Babe and the Cloak of Manliness', in *The Well Wrought Urn* (New York, 1947; London, 1968), 17–39.

play (which are indeed striking) concluded as though the significant image-train resolved the superficial strangeness. It does not, because it is not simply the individual images that are strange, but the very structure in which they emerge, and explanation must not dispel that.

Language has been under pressure from the beginning of the speech: here that is even more marked and more concentrated in syntax, words, and images; the argument is sustained, with the suppressed allusions to law and to religion emerging through what are in effect puns, though very far removed from jokes, and to them is now added a dimension of rhetorical grandeur which may have been implied before, but has never yet been allowed to develop. The primary sense of 'plead' here is the legal one, but associated with angels it immediately takes on the sense of 'beg' which leads towards 'pity' two lines later. The angels themselves appear first as a simple cliché metaphor, but immediately become concrete, blowing trumpets, and transpose through the babe into Heaven's cherubim; simultaneously the 'blast' is the jet of air on which the babe strides, and which can be seen as the jets blown through the trumpets, and it is also the sound which the trumpets make; in both senses it 'blows the horrid deed'.

The main image-trains are therefore of sight and of sound. The sound is an obvious crescendo from 'plead' through 'trumpet' to 'blast'. Sight is equally achieved as crescendo, first by becoming progressively more specific from the vague 'angels', through 'trumpet-tongued' into the insistently detailed 'naked new-born babe', and then by literal enlargement from 'babe' through cherubim, ambiguously represented either as cherubic babies or as androgynous adolescents in paintings of the period, which in final glory mount the sightless couriers of the air. 'Sightless' means both 'blind' (as winds move blindly) and 'invisible', and at this point the whole visual structure dissolves into the unvisualizable 'Shall blow the horrid deed in every eye | That tears shall drown the wind'. Sight and sound equally disappear in this rhythmic cadence, and distinct sense with them; the tears should be such as angels weep, but that sense is lost in the diminuendo which follows the climax; it is possible to rationalize the lines in various ways (wind causes eyes to water), but it is clearly irrelevant to do so: the grand vision has dissolved.

But Macbeth's speech does not end there, his argument is sustained in simple metaphor:

> I have no spur
> To prick the sides of my intent, but only
> Vaulting ambition ...   (25–7)

The whole passage is, of course, constructed out of metaphor, but in a peculiar way: the vision is a self-sustaining structure of words closely related by juxtaposition, but not by syntax. The repetition of 'like'—'plead like angels', 'pity, like a ... babe', extended by implication in 'or [like] Heaven's cherubim'—divides tenor and vehicle so emphatically that these should function as simple expository metaphors where all the stress is on the tenor, Macbeth's moral argument. In fact, we almost lost sight of that until it re-emerges in the final lines where the horse metaphor is indeed used again, but now it really is expository, affirming the nature of ambition. In the mean time the extraordinary baroque vision has been entirely created out of the metaphoric vehicles. The eruption stresses, certainly, the legal and (paramount here) the religious thoughts which Macbeth has tried all along to suppress, and thus far a Freudian account of the speech will satisfy; but only thus far. We certainly credit Macbeth with religious fears, but to credit him with the specific images of them would inevitably lead to Bradley's error of assuming him to be an exceptionally imaginative man. That would make him a poet, and depend on the same fallacy as believing that because most of Shakespeare's characters speak in blank verse they are all poets.

This constitutes an exceptional difficulty for the actor: not all the words he has to speak as Macbeth can properly be said to constitute part of his sub-text for the role. Macbeth and Richard III both demand virtuoso acting, but whereas Richard is a show-piece for the actor's skills, Macbeth offers virtually no scope for an actor's egotism. His language is, and is not, the property of his role, for it recurs in other speakers, most conspicuously Lady Macbeth. It is unique to the play, not to the man. Ben Jonson wrote in *Timber* 'Language most shows a man: speak that I may see thee',[1] and applied this principle in his plays, creating distinctive languages for his major characters; they often speak

---

[1] *Timber: or, Discoveries* ... (1641), in *Works*, ed. C. H. Herford, Percy and Evelyn Simpson, viii (Oxford, 1947), 625.

verse, but their poetry is an extension of their idiosyncratic speech, as with Mammon or Subtle, or commonly enough of the role they are for the moment playing, as with Doll, Face, or Volpone. There is almost no problem in Jonson's plays of characters speaking a language which is rather a property of the play than of themselves. Nor is there with some of Shakespeare's most obvious 'character' roles, such as Juliet's nurse or Shylock: their very distinctive speech intensifies into its own verse, but that verse remains distinctively their own and finds no reflection in other utterances (except by direct parody). That is, however, a fairly rare distinction in Shakespeare's work; minor members of the cast are always liable to break out into language of which, it is clearly understood, they can have no personal experience— the Welsh Captain, for instance, in *Richard II*, or Vernon in 1 *Henry IV*:

> All plumed like ostriches, that with the wind
> ⌈   ⌉
> Baiting like eagles having lately bathed,
> Glittering in golden coats like images ...   (4.1.98–101)

They *are* images: images of the show that Henry is putting on; and, because it is not simply show, images of a glorious honour which is one co-ordinate of the play; but they are not at all images of Vernon, whose language elsewhere is plain and undistinguished—we see the images and not the man.

But at least there the language is not in necessary contradiction of the man, an undefined soldier who may be supposed to rise to fine sights. Shakespeare sometimes went further:

> The west yet glimmers with some streaks of day.
> Now spurs the lated traveller apace
> To gain the timely inn ...   (3.3.5–7)

There has been wild speculation on the identity of the Third Murderer in *Macbeth*, but none about the First, who speaks these lines; they belong absolutely to the play and are alien to the speaker.

It is a large step from what can loosely be called 'choric' speeches for minor characters to the major soliloquies of the principals, for they are certainly men and women and unquestionably shown by their speech. But the mode of showing, and the degree, still varies substantially. Richard III's opening soliloquy is composed out of two much older conventions, that of the Presenter (or Chorus), and that of the self-declaration of the Vice; it becomes more than either in the knowing self-projection of Richard into both roles, yet it still serves both functions as well. Hamlet's soliloquies are usually discussed as direct self-revelations of a very self-conscious man, and that seems appropriate: he speaks a variety of languages, and when he waxes most poetical it is always recognizable as self-dramatization. All his languages do echo elsewhere in the play from speakers as diverse as Horatio, Polonius, Claudius, and the Ghost; the inner and outer bearings of man and environment are significantly equivalent, and just as the man has no easily recognized integrity, so also the play has no single language by which it can be identified. *Othello* is a different case, for what Wilson Knight dubbed the 'Othello music'[1] does substantially characterize the play; but it is spoken by Othello alone and so generates the well-worn critical problem that only Othello can articulate his own valediction: he alone has the language worthy of it—or, as T. S. Eliot suggested, he is 'cheering himself up'.[2] That is not a problem in *Antony and Cleopatra*, for although the play's characteristic magnificence is distinctively embodied in hero and heroine, their language is heard in various other mouths: in Philo's opening speech (1.1.1–13), Enobarbus's 'The barge she sat in' (2.2.197–246), and even from Caesar when he reminisces about Antony in the Alps (1.4.55–71). It adumbrates an imperial theme in the largest terms, yet it is always used by or about the man and woman: the principle that it 'shows' them is not violated. The fact that an actor may speak a language remote from the individual he represents is very clear here, yet it is not conspicuously a problem.

Macbeth is a different matter. We know the First Murderer speaks for the play, not for himself; so, really, do Duncan and Banquo in 1.6:

> DUNCAN
> This castle hath a pleasant seat, the air
> Nimbly and sweetly recommends itself

[1] G. Wilson Knight, *The Wheel of Fire* (1930), 107–31.
[2] T. S. Eliot, 'Shakespeare and the Stoicism of Seneca' (1927), in *Selected Essays* (1932), p. 130.

> Unto our gentle senses.
> BANQUO   This guest of summer,
> The temple-haunting martlet, does approve,
> By his loved mansionry, that the Heaven's breath
> Smells wooingly here ...   (1–6)

The benignity may fit Duncan; it is not especially characteristic of Banquo (nor necessarily inappropriate either). Ross and the Old Man in 2.4 function like the Welsh Captain, but with a language so like Macbeth's own that the distinction of man from words is even more striking:

> By th' clock 'tis day,
> And yet dark night strangles the travelling lamp;
> Is't night's predominance, or the day's shame,
> That darkness does the face of earth entomb
> When living light should kiss it?   (6–10)

The characteristics amplify as they describe Duncan's horses:

> Beauteous and swift, the minions of their race,
> Turned wild in nature, broke their stalls, flung out,
> Contending 'gainst obedience, as they would
> Make war with mankind.   (15–18)

That is not said about Macbeth, but it might perfectly be said by him. He sees strange sights, but they are not distinctively his. We do not credit all his words to his distinct consciousness, and frequently do not know whether to do so or not. An obvious case is his speech after the murder is revealed:

> Had I but died an hour before this chance,
> I had lived a blessèd time ...   (2.3.93–4)

The words so perfectly extend his private musings before the event that it is quite ambiguous whether they are here a private utterance or a public speech. Kenneth Muir argued that Macbeth was unconscious of the truth of his words, but he quoted Middleton Murry to the opposite effect: 'Macbeth must needs be conscious of the import of the words that come from him.'[1] The choice is vitally important for an actor, since it involves the

---

[1] See note on ll. 89–94 in his edn.

difference between an aside and an address to others on the stage. I am positive that Murry was wrong, but not positive that Muir was right. The ambiguity remains, and the actor need not eliminate it. But this is certain, that Macbeth may speak words beyond his consciousness; which means that his language may show us things other than the man.

I have shown that ambiguity in the opening soliloquy of 1.7; it does not only affect Macbeth: Lady Macbeth frequently uses a very similar language, and the actress has similar problems to contend with. They are made more acute by the more positive form of statement to which she is prone:

> I have given suck, and know
> How tender 'tis to love the babe that milks me ...
>
> (1.7.54–5)

The question whether Lady Macbeth ever had a child has been so much discussed that it is hard now to tell how odd her words should sound. In theory it should be impossible to go behind the direct assertion of a dramatic character and question its veracity unless the context gives her the lie. Macbeth does not retort 'When did you ever give suck?'; so there it is, she did. Or did she? Dover Wilson quoted Eckermann's *Conversations with Goethe* for 18 April 1827: 'Whether this be true or not does not appear; but the lady says it, and she must say it, in order to give emphasis to her speech.'[1] Exactly: and it is impenetrably ambiguous whether she means it, let alone whether it is true or not. But as an imaginative fact that babe is certainly very vivid to us in ways that are no part of Lady Macbeth's consciousness: it takes its place, with Macbeth's naked new-born babe, and all the other babes of the play, in a dimension well beyond the reach of the characters.

I do not believe that that example need be any special problem to the actress; but Lady Macbeth's earlier soliloquy, in 1.5, unquestionably is:

> Come, you spirits
> That tend on mortal thoughts, unsex me here,
> And fill me from the crown to the toe, top-full
> Of direst cruelty.
>
> (39–42)

The actress's problem is whether to make this a literal invocation of the spirits, in which case she must enact some appropriate

---

[1] Wilson, commentary note on 1.7.54.

ritual on the stage, or to project it as at least partly metaphoric, an extreme form of autosuggestion. Kenneth Muir argued decisively for the first, believing that it was Mrs Siddons's interpretation, quoting her 'Remarks on the Character of Lady Macbeth': '[She] having impiously delivered herself up to the excitements of hell ... is abandoned to the guidance of the demons she has invoked'.[1] But I am not certain this is quite literal: Mrs Siddons invoked her spirits in a whisper, and the effect is rather that the demons turned out more real than she imagined—as conscience does later in the sleep-walking scene (5.1). If she does literally enact a ritual, it becomes odd that no other such ritual ever occurs, just as the solitary reference to her child becomes odd if it is literally believed. Muir commented that we need not necessarily assume that Shakespeare himself believed in demoniacal possession; I agree, but would add that we need not necessarily believe that Lady Macbeth did either. If, on the other hand, the speech is allowed a primarily metaphoric force, then its extraordinary language tends to divide its reference (never, of course, precisely) between a relatively simple level corresponding to her consciousness, and a far more obscure level in which her words reverberate with images we do not specifically understand to be hers.

Lady Macbeth's speech therefore resembles Macbeth's of fifty lines later in its double focus—on herself, and far outside herself. It resembles it also in structure. She began, before the Messenger's entry, with a direct discussion of Macbeth's tricky conscience:

> What thou wouldst highly,
> That wouldst thou holily; wouldst not play false,
> And yet wouldst wrongly win.   (19–21)

a syntactic tangle which is closely echoed in Macbeth's tongue-twisting opening. From there, as Macbeth's images amplify through deep damnation to the naked new-born babe and Heaven's cherubim, so hers expand from

> The raven himself is hoarse
> That croaks the fatal entrance of Duncan
> Under my battlements.   (37–9)

---

[1] Muir, Introduction, pp. lviii–lix; Thomas Campbell, *Life of Mrs. Siddons*, 2 vols. (1834), ii. 12.

to 'Come, you spirits' and so to 'Make thick my blood' and

> Come to my woman's breasts
> And take my milk for gall, you murd'ring ministers,
> Wherever, in your sightless substances,
> You wait on nature's mischief. (46–9)

The spirits have become 'murd'ring ministers' and finally 'thick night' as the speech reaches its climax in Hell and Heaven:

> Come, thick night,
> And pall thee in the dunnest smoke of Hell,
> That my keen knife see not the wound it makes,
> Nor Heaven peep through the blanket of the dark
> To cry, 'Hold, hold'. (49–53)

The one verbal echo in Macbeth's speech is of the striking use of 'sightless' meaning both 'blind' and 'invisible'—'sightless substances' and 'sightless couriers of the air'. Finally, Lady Macbeth ends, as Macbeth does later, in plain language, though in both cases it is unclear whether the thought is complete or is interrupted: she ends 'To cry, "Hold, hold"' and he enters, while he ends 'which o'erleaps itself | And falls on th' other' and she enters. In both speeches the extraordinary imagery is left behind, and the thought-trains extend beyond it, so that the effect is as though the thought-trains represent the consciousness of the speakers while the image-train in its specific form has been raised above and beyond their distinct consciousnesses. Only, of course, in its specific form; they are understood to use metaphor, and to be able to see 'sights', but not necessarily to be aware of these specific sights.

We, on the other hand, are vividly aware of them, of the likeness between them, and of the likeness to other utterances by the Macbeths and by other people in the play—Duncan and Banquo, Ross and the Old Man, the First Murderer, and so on. The situation resembles that of a traditional mode of painting where human figures are shown at ground level in a rapt contemplation in which they may well see visions, but do not seem to be seeing the specific angels, devils, or other images which are painted around and above them. Pictorially that is a common enough convention; dramatically it may often be partially realized, but in such a fully developed form *Macbeth* is unique, as it is unique in the extent of specific visualizing it demands of its audience. The effect is achieved by an ambiguity of reference: Macbeth and Lady Macbeth do 'see', and they do use metaphor, yet the extraordinary visual details strike us as not actually theirs, as in fact a property of the play which exists outside them. Not merely outside them, but almost in opposition to them, as of something which it would be better for them if they *did* see: 'nothing is | But what is not' (1.3.142–3) has more meanings than Macbeth assigns to it, and 'what is not' eventually destroys him.

Their languages are finally brought together in 3.2. Lady Macbeth opens, as both of them had in Act 1, with a see-saw structure:

> Nought's had, all's spent,
> Where our desire is got without content;
> 'Tis safer to be that which we destroy,
> Than by destruction dwell in doubtful joy. (5–8)

She is interrupted by Macbeth's entry, and he seems to amplify her meaning:

> Better be with the dead,
> Whom we, to gain our peace, have sent to peace,
> Than on the torture of the mind to lie
> In restless ecstasy.   (21–4)

But their understandings are exactly opposite; she proposes to beat down the fear, 'Things without all remedy | Should be without regard—what's done, is done' (ll. 12–13); while he hints at further action:

> O, full of scorpions is my mind, dear wife—
> Thou know'st that Banquo and his Fleance lives. (39–40)

She meets that with ambiguity: 'But in them nature's copy's not eterne' (l. 41). She, it appears, understands 'they won't live for ever', whereas he takes her words to imply 'they can be killed': 'There's comfort yet, they are assailable' (l. 42); and so goes on to, 'there shall be done | A deed of dreadful note' (ll. 46–7). His meaning is patent, and she retreats from it, refusing rather than failing to grasp it, as he instantly recognizes:

LADY MACBETH   What's to be done?
MACBETH
> Be innocent of the knowledge, dearest chuck,
> Till thou applaud the deed ...   (47–9)

As he advances in ruthlessness, she retreats in fear; the roles established for them earlier in the play are reversed here, and he proceeds to develop an invocation that follows hers in 1.5 in form, word, and image so closely that in the theatre we need no special training to have at least the feeling that we have heard its like before:

>   Come, seeling night,
> Scarf up the tender eye of pitiful day,
> And with thy bloody and invisible hand
> Cancel and tear to pieces that great bond
> Which keeps me pale. Light thickens,
> And the crow makes wing to th' rooky wood;
> Good things of day begin to droop and drowse,
> Whiles night's black agents to their preys do rouse.   (49–56)

The connection is by no means confined to the opening phrase; 'scarf' recalls the 'blanket' of the dark; 'tender eye of pitiful day', the 'compunctious visitings of nature'; 'invisible', 'sightless'; 'light thickens', 'thick night'; 'crow', 'raven'; and, of course, the opposition of Night and Day is the foundation of the amplifying oppositions on which both speeches are constructed.

In one respect this speech does differ from its predecessors: they were soliloquies interrupted by the other's entrance; here both are on stage, and Macbeth concludes his thought in couplets which are superficially conclusive:

> —Thou marvell'st at my words; but hold thee still,
> Things bad begun make strong themselves by ill—   (57–8)

'marvell'st' is ambiguous; she might well marvel at his images, but at the plain level to which he has now returned, she is only said to be bewildered by what he is talking about—which is Banquo's murder; the point of their dialogue remains unmentioned and unmentionable, as Duncan's murder was in Act 1. This, as before, is the point at which Macbeth's thought-train runs out beyond the image-train, which here never seemed to suggest such literal invocation of spirits as Lady Macbeth's in 1.5.

What the couplet establishes is not an accord but the reverse, the final breakdown of communication between them. *Macbeth* is unique among Shakespeare's tragedies in centring on an intimate marriage (Othello's was never that). In Act I they did not hear each other's soliloquies, but always knew each other's thoughts. It is some while before Duncan's murder is made explicit between them, but they know at once that it is in each other's mind. That has, in the past, caused speculation about a missing scene or scenes in which they should have explained their thoughts to each other—which is absurd, because there is nothing even faintly unusual about this degree of understanding between any couple who live together (which is not to say that couples don't often find it remarkable, or that it does not function as part of the structure of supernatural hints in the play; only that it is as natural as sleep-walking). The characters of the two principals are simply and clearly defined, which has made them a favourite topic for junior exams; it is their relationship which is the focus of real interest in the play. It changes radically in 3.2, and they are never intimate again; simultaneously their roles are reversed, and he now displays the determination on blood which was once hers alone, but which she can no longer sustain.

This requires that they shall act very closely together; but the way in which the text presents it gives actors problems. Inevitably, seeking to enlarge the intimacy, they look for further expression of it. In the last fifty years or so this has tended towards sexuality, whether of the crudely obvious kind of Polanski's film (1971), or the more subtle embraces of Jonathan Pryce and Sinead Cusack (RSC 1986–7). The attempt is natural enough, but however it is done it seems curiously extraneous; it calls attention to an important fact: no play of Shakespeare's makes so little allusion to sex. There is none at all for the editor to explicate in the main scenes, allusions arise only in two places (besides the songs), the Porter's drunken bawdy in 2.3, and Lady Macduff's witty interchange with her son in 4.2. Both are striking interludes of 'normal' humanity offsetting the play's obsessive abnormality: the Porter ushers Macduff in to the hell of Duncan's murder; Lady Macduff's family domesticity gives way to slaughter. Where sexuality might most be expected, between the Macbeths, or in the Weïrd Sisters' obscenities, it is completely absent. There is another significant absence from the play:

although it is politically sensitive and perceptive about the hell of tyranny, it is so exclusively within the narrow society of the thanes; there is no sign of the populace, or of any concern for people at large. Even the laments for Scotland in 4.3, Macduff's 'Bleed, bleed, poor country' (l. 31) and Ross's speech at 164–73, treat the country as an emblem without specifying its people. The Weïrd Sisters, again, do not invite thought about the social problems of old women on their own, nor about witchcraft as an attack on women altogether, though the best known of Elizabethan studies of witchcraft, Scot's *Discovery of Witchcraft* (1584), concentrated on both, and castigated the witch-hunters for the evil manifestation they were; even King James, though he had been all too credulous in his early *Demonology*, became increasingly sceptical later, and though he never shared Scot's total scepticism, yet came to fancy himself more as an exposer of fraud than as a persecutor of witches.[1] Once again, it is the Porter alone who, in his brief scene, reminds the audience of the world that the play elsewhere so completely excludes; as he had from at least the fourteenth century onwards, the drunken clown inverted the assumed hierarchical order of things to assert human concerns supposed to be suppressed; his role and his language, coarse and bawdy prose, are indivisible, and the derisive laughter he so unexpectedly secures (when well played) radically adjusts our perspective on the play, illuminating the exclusions which otherwise we could only suspect. His role as devil-porter relies on the Harrowing of Hell in the mystery-play cycles, though he abandons it before he admits Macduff:[2] however effective it may be to identify Macbeth's castle with Hell, it is not possible to see Macduff as Christ.

The language that I have analysed as characteristic of the play is extraordinarily rich in what it *does* develop, and remarkable too in its exclusiveness; in politics, *Macbeth* contrasts strikingly with *Julius Caesar* before it and with *Coriolanus* after; they both offer Rome as an entire city-state, and *Lear* indicates a whole society through the curious range of Edgar's disguises; *Antony and Cleopatra*, probably written in the same year as *Macbeth*, is as comprehensive in its political concerns as it is insistent on sexuality. The contrasts identify the peculiar concentration of this

---

[1] See pp. 78–9.
[2] See pp. 79–81.

play, represented not only in its brevity overall, but also in the density of language which makes commentary so difficult and so rewarding. But this language is closely related to another that recurs in the play, where apparent density turns out only to be tortuous courtesy, where commentary is laborious and un-rewarding. The bleeding Sergeant's inflated rhetoric in 1.2 has the function of the classical messenger to explain its diffusion, but Ross's obscurity in 1.3 is more typical:

> The King hath happily received, Macbeth,
> The news of thy success; and when he reads
> Thy personal venture in the rebels' fight,
> His wonders and his praises do contend
> Which should be thine, or his. (89–93)

Superficially this may resemble Macbeth's language, but in fact it is its opposite, and the difference is felt when Macbeth speaks aside a few lines later:

> This supernatural soliciting
> Cannot be ill, cannot be good. If ill,
> Why hath it given me earnest of success,
> Commencing in a truth? I am Thane of Cawdor.
> If good, why do I yield to that suggestion
> Whose horrid image doth unfix my hair
> And make my seated heart knock at my ribs
> Against the use of nature? (131–8)

Macbeth's words amplify through horrid imaginings to murder, and he concludes that 'nothing is | But what is not' (142–3). The whole process, however diffuse and rhetorical it may superficially appear, is at once logically clear and expressively condensed. Ross's speech was neither; his language is an elaborate mask which conceals no substantial meaning.

Later in the play, however, the mask does develop a function, when the conditions of tyranny deprive any communication of mere courtesy, and meaning must be obscured because no man can be trusted. This characterizes 3.6, when Lennox and another lord fence verbally with each other before their mutual sympathy is established:

> How it did grieve Macbeth! Did he not straight
> In pious rage the two delinquents tear,

That were the slaves of drink, and thralls of sleep?
Was not that nobly done? Ay, and wisely too—
For 'twould have angered any heart alive
To hear the men deny't. So that I say,
He has borne all things well ...   (11–17)

A similar double-talk baffles Macduff when Malcolm receives him
in England in 4.3, and is put to a different purpose as Ross delays
telling of the murders of Macduff's wife and family.

The primary language of *Macbeth*, that of the thanes, divides
into three distinctive usages: the empty elaboration of courtesy,
the masked talk of potential conspirators, and, above all, the
condensed approach to what is not that is most distinctive of the
play. In total contrast is the Porter's prose; and another contrast,
equally strong, is provided by the Weïrd Sisters' short-lined verse,
variable in rhythm and in rhyme, elliptical and enigmatic in
sense. They do not joke, and they are not bawdy, yet their
weirdness has a comic edge to it which intensifies the sinister
malice they engender. It was probably a particular tradition in
their performances before the Commonwealth which led to their
being still played by men after the advent of actresses, and they
continued to be the province of clowns until the nineteenth
century. The eighteenth century came to question the pertinence
of their antics, and Kemble restrained what he found an affront to
the play's dignity—*his* dignity as the hero. No doubt eighteenth-
century 'Sisters' got out of hand, but male actors still sometimes
take one or more of the parts, and I should record that a notably
camp performance by the Second Witch at Stratford, Ontario, in
1971 was more effectively mysterious than any other I recall. It is
not only in appearance that the Sisters present puzzles for performance, and any account of their language should recognize this.

3. *Baroque Drama*

Critical discussion of *Macbeth* throughout the eighteenth and
nineteenth centuries concentrated on the characters of the hero
and his lady. Initially the problem was seen as essentially moral,
but in the nineteenth century attention shifted decisively towards
psychology, culminating in Bradley's celebrated essays in 1904.[1]
It was a central text for Bradley's theory of tragedy, but he

[1] A. C. Bradley, *Shakespearean Tragedy* (1904).

recognized, in his opening paragraphs, the exceptional importance of imagery in this play, and it is not surprising that in the twentieth century it became a central text for imagist criticism, opening with L. C. Knights's celebrated attack on Bradley, *How Many Children Had Lady Macbeth?*[1] The patterns of iterative imagery were seen as developing psychological, moral, or spiritual patterns in the work of Knights, Wilson Knight, Cleanth Brooks,[2] and others; but the synthesis did not satisfy, and there emerged an ill-defined conflation of the two critical traditions— character and imagery—which seems to have established it as the clearest, most teachable, of Shakespeare's tragedies. The obvious problems of sensational credulity—witchcraft, ghost, apparitions—are not discussed, or are assigned to projections of Macbeth's fevered brain, or to the atmospherics of Evil, or—worse—to the supposed credulity of Jacobean audiences.

Readers, however critical, are always selective, but a theatre audience is obliged to attend to whatever it is shown: it may sleep, but it cannot skip. The strangeness of *Macbeth*, all that makes it alien to a simplistic tragic tradition, can only be ignored in the theatre if it is literally cut. Its dramatic form, I have argued, depends upon a spectrum of dramatic illusions which are set in unusually sharp contrast with the naturalism invested in the Macbeths' relationship. The result is a sharp dichotomy between realism and supernatural phenomena. Neither is viewed simply: the Macbeths are presented to us in two ways, firstly by others, and secondly by their own appearances on stage, whether singly or together, or in asides (which are unusually numerous in this play). To others, they are at first patriotic hero and gracious lady, and at last, bloody butcher and fiend-like queen; as we see them, neither description seems particularly appropriate, despite their deeds, for the subtlety of their presentation militates against such hyperbolic language. The supernatural is also seen ambivalently: there is, certainly, an invitation to credulity, or at least to what Coleridge called 'the willing suspension of disbelief'; but there is also an inclination towards scepticism. The movement towards rational explanation in the last act allows the antithetic structure

---

[1] Cambridge, 1933, reprinted in *Explorations* (1946).
[2] L. C. Knights, *Some Shakespearian Themes* (1959); G. Wilson Knight, *The Wheel of Fire* (1930), *The Imperial Theme* (1931); Cleanth Brooks, *The Well Wrought Urn* (New York, 1947; London, 1968).

to hint at (but certainly not to articulate) the possibility of rational explanation for supernatural phenomena themselves. These are precisely the terms that I would use about the aesthetic structures of Roman baroque art in the first half of the seventeenth century, which can provide a context for understanding *Macbeth*.

Baroque art is conventionally assumed to originate in Rome, in the work of Carracci and Caravaggio, who both (by coincidence) arrived there in the early 1590s. Both were dead by 1610. The dating is convenient to my argument, but I must stress that there is no evidence whatever that anything was known in London of their work before Inigo Jones's expedition with the Earl of Arundel in 1614–15.[1] My argument is entirely by analogy, between dramatic art in London and visual art in Rome. But analogy is enlightening here, because of the difference in apprehension between the two arts: dramatic art is strong in narrative action, but perception of whole structures is difficult because perception is sequential and the whole can only be grasped in memory; whereas visual arts can only suggest action, but the structure is immediately perceived in its entirety.

Caravaggio is often associated with crude realism, as if his treatment of miracles were only a cynical response to ecclesiastical patronage. His Conversion of Saul makes its initial impact with the aggressive rump of a foreshortened cart-horse; the naturalism is sustained by the old peasant at the horse's head. The painting has a double composition: the first is square, parallel to the frame, governed by horse and groom across the top and down the right side, and extended across the bottom by the brilliant red cloak into the left-hand corner. For all its foreshortenings it remains two-dimensional. It is only at the bottom that Saul's head is perceived—but this introduces the second, radically different, composition which focuses entirely on Saul and the decisive perspective in which he is illuminated by a dramatically unnatural light below the horse—and for this the first composition, horse and all, becomes merely a dark frame to the three-dimensional illusion.

That amounts to an illusion trick similar to the well-known

---

[1] Described in John Summerson, *Inigo Jones* (1966), 35–7.

1. *The Conversion of Saul*, by Caravaggio

shaded cubes which are perceived either as receding from you, or as projecting towards you, but cannot be seen both ways simultaneously. Here the realistic horse is seen first, Saul second; but again the one perception excludes the other. The transition from one to the other dramatizes the concept of miracle, which is here by no means a squashed-out gesture to the Church: if you cut out the lower half of the painting the rest loses all its impact and becomes merely the sentimental study of horse and groom that it has been in genre imitations ever since. So, here, optical illusion is used to establish a concept of miracle in sharp contrast with the initial naturalism; illusion, in other words, mediates between natural and supernatural.

Caravaggio used a similar device in his Martyrdom of St Matthew, but here the double vision makes a kind of visual pun. The centre is dominated by the brutal and disturbingly sensual figure of the almost naked executioner; with the other two naked figures below he forms a solid triangle that excludes both the dandies to the left and the clouds to the right. St Matthew is (like Saul) less immediately conspicuous, lying across the middle with his hand raised in a futile gesture of self-protection. But that hand connects in quite another way with a very different nakedness among the clouds—the seductive body of the androgynous angel leaning over a cloud to offer the martyr's palm. And in that switch from supplication to acceptance the composition changes from death to resurrection.

Two generative works for baroque aesthetic theory seem to have been St Ignatius's *Spiritual Exercises* (1548) and St Teresa's *Life* (1562). The two saints shared the distinction of unusually rapid canonization, and they also shared the careful deployment of very matter-of-fact prose, emphasizing natural experience, for the discussion of highly sensational material, proposing supernatural experience.

St Ignatius's writing is so restrained that its sensational impact throughout Europe is almost surprising; but sensational it literally is. The exercitant is directed to select a scene from the New Testament and to subject it to systematic sensory contemplation. The first act is to visualize the scene as precisely as possible *as it was*; that is to say, realistically in every detail. There then follows, on successive days, the deliberate application of each sense in turn to that scene. Ignatius only specifies one such scene,

2. *The Martyrdom of St Matthew*, by Caravaggio

very briefly, a contemplation of Hell: you will see flames, hear groans and screams, feel unbearable heat, taste ashes, smell brimstone—and do each one separately with such intense concentration that the result is necessarily a progressive amplification. On the final day the sensations of the previous five are revived and combined in a final synaesthesia, in which the original scene of day one must indeed have been transformed.

St Ignatius's *Exercises* are certainly sensational; they are also to be very deliberately controlled: at frequent junctures the exercitant must pray for guidance and profit. The experience is to be at once self-consuming and self-aware. St Teresa's work is necessarily different, because her experiences were involuntary: her matter-of-factness was aimed at allaying the suspicions of the Holy Office, so she presents herself as the busy administrator of her rapidly expanding Order, who found levitation a tiresome distraction from her desk (which remained below). But she is equally direct about the sensuality of the experience when an angel drove a spear into her heart, which she said was emphatically of the body, and yet of the spirit too, a pain which was also an exquisite pleasure.

Teresa was describing miracles, Ignatius was not. The similarity is obvious, both in the sensory consciousness and in the rational awareness; but the essential difference made her a *subject* for an art whose *methodology* was based on him. Bernini told his son that he carried the *Spiritual Exercises* with him at all times, and his best-known work is the Ecstasy of St Teresa. Bernini used the full resources of architecture, sculpture, stucco, and paint to manipulate the viewer, first into the right physical position for contemplation, and then through the stages of an Ignatian exercise: you begin with the realistic Cornaro cardinals in their stage boxes, who first call your attention to the chapel from the nave of the church and then (when you enter) turn out to be, wittily, a wholly inattentive audience, unaware of the drama above the altar; but the seductive and sardonic angel redirects your attention to St Teresa herself. The cardinals are carved in unpolished stone, the angel in partly polished marble—the highest polish is reserved for the saint on whom contemplation must dwell. She reflects the sole source of light at this level which streams down the gilt rods from a window concealed behind the architrave. Bernini contrived the effects of artificial light in a dark

3. *The Cornaro Chapel, S. Maria della Vittoria, Rome,* by Bernini. An eighteenth-century painting of the chapel in its original state, now in the Schwerin Museum, is reproduced in H. Hibbard, *Bernini* (1965), 129.

4. *The Ecstasy of St Teresa*, by Bernini: central group

5. *The Ecstasy of St Teresa*, by Bernini: Teresa's face

theatre from a natural source so that the illusion of mystery is as clear as it is dramatically effective. The sculpted figure becomes (like Bernini's pagan works in the Villa Borghese) a study in metamorphosis—between stone and flesh—and the intense sensuality finally focuses on her face, the most perfect representation in art of a woman in orgasm; an orgasm which further metamorphoses flesh into divine love, Teresa's reception by Christ which is the terminal point of the spiritual exercise represented on the remote ceiling in paint on stucco—or so it was until the paint was allowed to fade and flake, and the clear window which once illuminated it brilliantly was filled with chocolate glass a hundred or so years ago.

Bernini turned a shallow bay into a chapel, and the chapel into a theatre: the whole is an elaborate exercise in visual illusion that also evokes the sense of touch (precisely controlled, for the marble flesh, though close, is out of reach). But it was decidedly not mere illusionism (as late seventeenth-century baroque commonly was): the means are all exposed, and the shifts of genre through architecture, varying modes of sculpture, and paint preclude naïve response and insist on consciousness of the process itself. Optical illusion was, for baroque art—for Bernini as much as for Caravaggio—not merely a lavish technical display, but a controlled exercise in the study of illusory experience: at once the most sensational and the most rational of aesthetic forms. It was the fantastic art of an insistently rationalist age, dramatizing the achievements of Galileo, Descartes, and later, Newton, whose achievements in systematic thought in general and in the study of optics in particular led in each case towards a rational theology. The seventeenth century often baffles in its efforts to reconcile scepticism and credulity without compromising either: to mediate between them was precisely the function of illusion in baroque aesthetics. Visual or dramatic illusions must, in themselves, always be capable of explanation as rational as Birnam Wood. But their appeal is to amazement and, in so far as the explanation is only partly understood, they provide a suggestive analogy for supernatural phenomena, hinting at the possibility of rational comprehension of them equally. Hence the necessity of a basis in realism, and on making the techniques open to inspection.

Jacobean drama was, by law, strictly secular, so the 'miracles' it offered were less committed to credence than Catholic art in

Rome; I have stressed the scepticism, or at least the rationalism, that Shakespeare carefully establishes in his spectrum of illusions, and *Macbeth* certainly displays its techniques. They rely on well-established conventions for the open theatres and would be relatively easily accepted, but there is no pretence that they are not devices. That certainly does not negate their effectiveness, and the play makes its perception of evil very powerful. Like the baroque artists, Shakespeare displays a supremely accomplished art to sustain at once credulous amazement and rational intelligence. There is no reason to suppose that he knew Ignatius's work at first hand (though doubtless there were copies in Catholic households), but Lady Macbeth's invocation to the spirits of darkness takes the form of an Ignatian exercise inverted; and though her language is like her husband's, it is distinguished by the fact that her experience is voluntarily induced, whereas his, in this respect like Teresa's, is involuntary and unwelcome. In its carefully deployed spectrum of dramatic illusions, the play offers a thorough analysis of the epistemological relation between belief and knowledge. In that way it is profoundly serious, but it is radically different from *King Lear*; contrasting the two, a friend of mine remarked that going to see *Macbeth* is (or ought to be) fun, which *Lear* certainly is not. Illusion, however powerfully effective, is entertaining, and the intelligence displayed allows some detachment from the disturbing perceptions created; the Weïrd Sisters are not comics, but there is wit in their rough magic. Equally, there is wit in the ironic intelligence of Caravaggio's and Bernini's work, as well as magnificence.

Baroque amazement, however, had also a strong political potential. Caravaggio's revolutionary inventions were (and are) seen as purveyors of social revolution, and were several times rejected by the patrons who had commissioned them; Bernini's work was intellectually responsible, but devoted to the support of the divine monarchy of the absolutist popes of the seventeenth century,[1] and if it satisfied the intellectual speculations of an educated class, it seems also calculated to impose awe on the uneducated. At the end of the century his imitators turned it

---

[1] His wish to serve Louis XIV as well was less successful: at the age of 66 he accepted an invitation to Paris; he completed a bust of the *roi soleil*, but none of his successive designs for the Louvre was built. The visit, which was richly documented, is discussed in H. Hibbard, *Bernini* (1965), 168–83.

6. *The Apotheosis of James I*, by Rubens

exclusively into an instrument of social control, aimed at mere illusion to invoke wonder and awe. *Macbeth* is known for its apparent gestures to Stuart absolutism, and James I's affirmation of divine right; but it is equivocal about that monarchy (see pp. 72–6): Banquo's behaviour is ambiguous (3.1.1–39); Duncan, though 'good', is not holy; St Edward the Confessor is restricted to legendary report, and his disciple, Malcolm, is a very dubious hero. The potential is recognizable, but critically so. The situation was different a few years later when Rubens was hired by Charles I to make of his father, James I, the first Christian monarch accorded apotheosis in the centre-piece of the painted ceiling he added to Inigo Jones's Banqueting House in Whitehall, formerly the baroque theatre of court masques. Rubens painted James's face with striking naturalism, warts and all, drunken intelligence surprised and frightened by levitation and by the ample female sensuality that sustains him. The composition is based on the illusory perspective of a hole in the ceiling through which the King is being wafted to Heaven, but there is no pretence of credible illusion as it transforms realism into exotic fantasy. The two interact far more seriously in *Macbeth*: here, as in Caravaggio and Bernini, illusion mediates between natural and supernatural.

4. *Staging*

My contention is, therefore, that *Macbeth* is in all significant respects baroque drama. But it was not conceived for a baroque theatre, whose primary condition was darkness, which could be lit to make effective a complex machinery for the transformation of visual scenery, and be further enhanced by music, song, and dance. For St Teresa's ecstasy Bernini combined architecture, sculpture, and painting; the seventeenth century came to believe that the theatre could combine all these with music as well, a synaesthesia of the arts. So its characteristic invention was opera. In the Restoration, *Macbeth* became an opera; but it began as a *tour de force* for the daylight theatre, with limited mechanical aids. Two or three years later, with the purchase of the Blackfriars Theatre, the company acquired a naturally dark theatre. It was not, so far as we know, capable of much of the baroque machinery which Inigo Jones set up for the court masques, with magic transformations and all, nor could it be developed very far in that direction while the same repertory had to be performed in both theatres, and so had to transfer without drastic alterations. But the technique of illusion on which *Macbeth* depended in such extreme form must have been an embarrassment at the Blackfriars, for darkness is no miracle in a dark theatre: the play needed revision to compensate for what it lost, and to take advantage of the opportunities offered.

This is guesswork, based on the text we have, which appears to have been revised to include the Hecate scenes, 3.5 and 4.1 (see pp. 51–5), spectacular shows involving singing and dancing and a goddess ascending into the heavens with her attendant spirits; in fact, the material of the masques. Shakespeare's last plays, *Cymbeline*, *The Winter's Tale*, and *The Tempest*, all have such scenes, and at least some of them seem to have used music and choreography composed for the court; it has frequently been conjectured that the antimasque of witches in Jonson's *Masque of Queens*, for which the King's Men had been employed, might have been pressed into service in *Macbeth* for the final dance in 4.1 (see Appendix B). Whether that was so or not, the operatic material added in these two scenes must have helped substantially to enhance the climax of the Witches' scenes, and of the play's spectacular illusions, and it could have been performed in the Globe as well as in the Blackfriars. Probably there were other modifications made at the same time, though where and what cannot clearly be detected (p. 57). *Cymbeline* has a descending god for Posthumus's dream in 5.5; *The Tempest* is, overall, close to masque in form, but has also a specific masque with the goddess Juno descending to join Iris and Ceres. *The Tempest* is generally regarded as a baroque play, but controversy has raged as to whether it should be performed with minimal mechanical effects, or should make full use of the resources of modern theatres; the truth surely is that it is an astonishing achievement of theatrical technique because it is equally effective in either state—conceived to be played at both the Globe and the Blackfriars.

That, I suggest, was the purpose of altering *Macbeth* quite early after its first production. But if the original performances were deprived of lighting effects, they were not without sound. Frances Shirley, in *Shakespeare's Use of Off-stage Sounds*,[1] points out that

---

[1] Lincoln, Nebr. 1963; pp. 168–89 discuss *Macbeth*.

there were more calls for music in the Folio stage directions for *Macbeth* than for any other of the tragedies. She also believes that all other relevant noises were literally heard in the theatre, e.g. horses' hooves, the mew, croak, and whine of the witches' familiars, the owl's screech, the bell, and more certainly thunder. Music in the Globe consisted of trumpets and consorts of viols or hautboys, usually in groups of four each. Trumpets were without valves and so, like modern bugles, were restricted to simple calls ('Sennets', 'Flourishes', etc.) for military or ceremonial action; drums accompany marches. Viols were used for soft or soothing music and are not indicated in this play, but hautboys are, to accompany the feast in Act I (not in 3.4), and for the show of kings in 4.1. Hautboys, ancestors of the modern oboe, provided 'loud music' and, like the viols, were of various sizes, usually alto, tenor, and bass. There is no indication what instruments accompanied Hecate in 3.5 and 4.1.

There is an eyewitness account of *Macbeth* performed at the Globe in 1610 or 1611, in Simon Forman's notes for a projected book on plays 'for Common Policy', i.e. public morals. Such records are very rare and it is bitterly disappointing that Forman's recollections of the performance seem to have been mixed with a reading of Holinshed, and his visual impressions eked out from the illustrations. The passage is printed in full and discussed in Appendix C. In any case, though he is fairly detailed on the early part of the play, he recalls very little after Banquo's ghost in 3.4 except for the sleep-walking. Despite his interest in witchcraft (after his death he was implicated in the crimes of Lady Essex, see below, p. 64), he makes no mention of the apparition scene in 4.1, let alone of Hecate, so we cannot tell which version of the play he saw.

The introduction of musical spectacle left a certain awkwardness in producing material radically unlike anything else in the play so late on in the performance, and at some later date (just possibly before the Commonwealth, but probably after) another musical scene was added for the witches after Act 2 (see Appendix B). Further development was natural as theatres developed more elaborate resources, though none is documented before the Restoration. *Macbeth* was listed among the plays allocated to the Duke of York's company when new theatres were licensed in December 1660; the company was run by William Davenant who had worked with the King's Men in the 1630s, and when he returned from France in the 1650s he experimented with opera as a public entertainment. The form which he subsequently developed was of a spoken verse drama with substantial musical interludes involving some at least of the main characters. In Dryden and Purcell's *King Arthur* (1684), the musical parts centre on the witches who also control the main action—a structure very likely imitated from *Macbeth* as Davenant presented it. The play was licensed for revival in November 1663, but the first performance on record was a year later, when Pepys commented fairly laconically in his diary for 5 November 'to the Duke's house to a play, *Macbeth*, a pretty good play, but admirably acted'. Two years later, in December 1666, his interest had quickened: 'saw *Macbeth* most excellently acted, and a most excellent play for variety', and ten days later he commented more fully: 'which, though I saw it lately, yet appears a most excellent play in all respects, but especially in divertisement, though it be a deep tragedy; which is a strange perfection in a tragedy, it being most proper here, and suitable'. He saw it yet again on 19 April 1667, with equal enthusiasm, and this time is more explicit about the 'variety of dancing and music'.[1] Pepys's comments are particularly interesting because he stresses his surprise at the combination of tragedy with 'variety' and makes it clear that they worked together.

It sounds as though a change had come over the play between November 1664 and December 1666. The theatres were closed during the great plague of 1665–6, and it may have been then that Davenant undertook the full-scale rewriting of the play as an opera. Davenant died in 1668, and his company was continued under his widow and later his son. They built a new theatre in Dorset Gardens which opened in November 1671, and revived *Macbeth* there early in 1672. John Downes, who had worked for Davenant as prompter (stage-fright ended his acting career at his first attempt) from the early 1660s, wrote his memoirs shortly after his retirement in 1706. He mentions *Macbeth* for the first time in 1672, but does not say it was new; in fact, his remarks suggest rather a lavish fresh production of existing material:

The tragedy of *Macbeth*, altered by Sir William Davenant; being dressed in all its finery, as new clothes, new scenes, machines, as flyings for the

---

[1] *The Diary of Samuel Pepys*, ed. Robert Latham and William Matthews, vol. v, *1664* (1971), 314; vol. vii, *1666* (1972), 423; vol viii, *1667* (1974), 7, 171.

witches; with all the singing and dancing in it: the first composed by Mr. Lock, the other by Mr. Channell and Mr. Joseph Preist; it being all excellently performed, being in the nature of an opera, it recompensed double the expense; it proves still a lasting play.[1]

Downes's rich memory was not always reliable, nor his language precise: he was doubtless right that costumes, scenery, and machines had been newly designed, but it is not clear whether Locke's music was new then, or had been composed in Davenant's lifetime. Locke's name continued to be associated with the music used in the theatre into the present century, though it seems that this was actually composed by Richard Leveridge for a revival at Drury Lane in 1702. Leveridge was a bass singer who played Hecate himself on and off until the 1750s, and seems to have been content to see his work later attributed to his more illustrious predecessor. In the late eighteenth century, Purcell's name was sometimes given, but even in 1672 Purcell was only 13, so that was always impossible.[2]

Davenant did not merely add operatic interludes;[3] although he respected a great deal of the original text, he recast the play in several important respects. The language is modified to suit the demand for a plainer diction, represented by the easy conversation of a gentleman, and gentry as such appear on the stage, where there were none in Shakespeare's text: both the Porter and the Doctor disappear altogether, but their necessary functions are now performed by Seaton, who has risen socially and so must be redeemed by a belated change of sides in Act 5. The play conforms, in short, to the norms expected by Society in the London season; Macbeth himself shows decent restraint even towards the end, when his original 'The Devil damn thee black, thou cream-faced loon: | Where got'st thou that goose-look?'

---

[1] John Downes, *Roscius Anglicanus* (1708) ed. John Loftis (Los Angeles, 1969), 33.

[2] For the complicated history of the music used in the 18th cent. I am entirely indebted to Dr. Christopher Field of St Andrews University. Leveridge's name seems to have been deliberately suppressed: it appears in the Fitzwilliam Museum, Cambridge, MS Mus. 87, but there is a hole in the title-page of British Library Egerton MS 2957, the copy which William Boyce used for his edition in 1770, where he attributed it to Locke. That was the basis for numerous editions in the 18th, 19th and 20th centuries of 'Locke's music'. See R. E. Moore, 'The Music to *Macbeth*', *Musical Quarterly*, 47 (1961), 22–40; R. Fiske, 'The *Macbeth* Music', *Music and Letters*, 45 (1964), 114–25; R. Fiske, *English Theatre Music in the Eighteenth Century* (1973), 25–9.   [3] See Appendix B.

(5.3.11–12) becomes merely 'Now friend, what means thy change of countenance?' By the same token such rhetorical figures as the naked new-born babe, Heaven's cherubim, and the bear tied to a stake disappear, and care no longer has a ravelled sleeve.

Davenant's attention was not simply to taste, but to larger issues of social and political significance. His major structural alteration was aimed to balance the forces of evil and good by developing Macduff into a central hero, and giving his wife a generally 'correct' feminine role in direct contrast with Lady Macbeth (they first appear together). The diabolism latent in the original text loses almost all its force: Hell is scarcely mentioned, and the witches appear to the Macduffs, as ready to support them as they are to destroy Macbeth; the supernatural is reduced to exciting theatrical effect, the equivocal interest in it eliminated in favour of the polite rationalism of late seventeenth-century theology. Macduff's expansion to heroic stature is partly achieved by transferring to him much of Malcolm's role, which leaves Malcolm even more diminished than before, but no longer an ambiguous figure, rather he is a minor one who can finally be given the symbolic role of a constitutional monarch; his last speech is severely pruned and Macduff has an entirely new valediction:

> So may kind Fortune Crown your Reign with Peace,
> As it has crown'd your Armies with Success;
> And may the People's Prayers still wait on you,
> As all their curses did Macbeth pursue:
> His Vice shall make your Virtue shine more Bright,
> As a Fair Day succeeds a Stormy Night.

The capitals pick out effectively the tone and the values which Davenant's version asserted. The transformation is radical, but it is also remarkably skilful and tactful, and in no way deserves the contempt with which it is usually treated.[1] The process of converting *Macbeth* for baroque theatre is complete here, but it is completed in the manner of late seventeenth-century baroque, without the complex ambiguity of Caravaggio's, Shakespeare's, or even of Bernini's work.

---

[1] See *Davenant's Macbeth from the Yale Manuscript*, ed. Christopher Spencer (New Haven, 1961), 1–16.

7. Macbeth with Witches, show of kings, and Banquo's Ghost: frontispiece to Rowe's edition, 1709, presumably based on Betterton's performances

Davenant's opera, for all its expansion into heroic drama, still made Macbeth himself the central figure, and he was played by the great Thomas Betterton; when Betterton was ill for several months in 1667, Pepys was seriously disappointed, and his wife once walked out.[1] In the eighteenth century, the operatic additions remained, but the dramatic adaptation largely disappeared, and the text reverted to something closer to the original. The heroic idea lost conviction, and Macduff shrank back to his former dimensions, but Malcolm did not expand. Historical interest in the text and in Elizabethan language encouraged a return towards the Folio text from the time of Rowe (1709), Pope, and Theobald to the central achievement of Samuel Johnson in

[1] Pepys, viii. 482–3, 521.

40

his *Dictionary* (1755) and in his edition of Shakespeare (1765). For him, the play was 'deservedly celebrated for the propriety of its fictions, and solemnity, grandeur, and variety of its action; but it has no nice discriminations of character, the events are too great to admit the influence of particular dispositions, and the course of the action necessarily determines the conduct of the agents'.[1] The Enlightenment evidently found the witches too improbable for serious drama, but Johnson defended Shakespeare with a learned essay on Jacobean credulity. The play, though still performed in contemporary dress, was no longer seen as strictly contemporary, but the supernatural was becoming very popular fictionally in the new genre of gothic novels, and *Macbeth* was a fertile source for Horace Walpole, Mrs Radcliffe, and numerous others.

Johnson claimed that 'Lady Macbeth is merely detested; and though the courage of Macbeth preserves some esteem, yet every reader rejoices at his fall'.[2] Shakespeare did not allow his actor (Burbage) to share this reaction with his audience, for he gave him no dying words; Betterton was allowed one line in an ending which mainly celebrated Macduff, but Johnson's pupil Garrick, the next great actor to play the role, provided himself with a lengthy speech, at once penitent and self-damning: the actor could therefore align himself with the audience's judgement, in a fictionally medieval world. For all Garrick's restoration of Shakespearian language, readers (of expensive scholarly editions) and audiences were moving further apart, and popular texts were based on theatre prompt-books, often sold in the foyer at performances. Shorn of political and social interest (though the Doctor returned as the medical profession rose in social esteem), the play focused more and more on the conflict of the protagonists, Macbeth and Lady Macbeth. Garrick was praised for the naturalism of his acting, but his Lady, Mrs Pritchard, seems to have dominated him on stage, and made the more lasting impression.

The later eighteenth century was preoccupied with psychology; Garrick's concentration was not solely an actor's egotism, and at the end of the century John Philip Kemble developed the interest further: he did not bring Banquo's ghost on the stage, and so made of him a figment of Macbeth's psychic disturbance.

---

[1] 'General Observation', reprinted in *Dr. Johnson on Shakespeare*, ed. W. K. Wimsatt, Penguin Shakespeare Library (1969), 133.　　[2] Ibid. p. 134.

8. Garrick and Mrs Pritchard in 2.2

9. Sarah Siddons in 5.1, Covent Garden, 1812

Popular demand from all parts of the house restored the Ghost,[1] but *Macbeth* was (like *Hamlet*, though less fully) turning in on the hero's self. It is not surprising that it was, more than ever, Lady Macbeth who dominated: Kemble's sister, Sarah Siddons, found more nuances in the role than Johnson had allowed, and became the most celebrated of all actors in it, at once psychologically subtle and intensely moving. It became a study of her version of being a woman, and an interesting one; a contemporary described her performance as 'womanly' and she included in this an independence like her own. Mid-nineteenth-century Lady Macbeths found themselves circumscribed by a concept of femininity which included subservience, like David Copperfield's wives. At the end of the century, Ellen Terry 'wheedled' compliance from Irving's officer and gentleman.[2]

Concentration on the central figures, however much it interested critics of the time, is none the less misleading about the performances. Kemble opened his rebuilt Drury Lane with *Macbeth* in 1794; the new theatre was huge, and the audience was very remote (as in Covent Garden still). For Mrs Inchbald, introducing his text (the shortest yet) in Longman's prompt-book series of 1808, the play was still 'this grand tragic opera' in which she found combined 'that which is terrific, sublime, infernal ... supernatural agency is produced and conducted by such natural means, that spectators return again to their childish credulity, and tremble, as in the nursery, at a witch and a goblin'. Kemble followed Charles Macklin in giving the play historical (i.e. gothic) sets and 'Scottish' costumes:

The huge rocks, the enormous caverns, and blasted heaths of Scotland, in the scenery;—the highland warrior's dress, of centuries past, worn by the soldiers and their generals;—the splendid robes and banquet at the royal court held at Fores;—the awful, yet inspiring music, which accompanies words assimilated to each sound;—and, above all—the fear, the terror, the remorse;—the agonising throbs and throes, which speak in looks, whispers, sudden starts, and writhings by Kemble and Mrs. Siddons, all tending to one great precept—*Thou shalt not murder*—render

---

[1] A. C. Sprague, *Shakespeare and the Actors* (Cambridge, Mass., 1945), 256–7 and n. 106.

[2] Marvin Rosenberg, 'Macbeth and Lady Macbeth in the Eighteenth and Nineteenth Centuries', in John Russell Brown, ed., *Focus on Macbeth* (1982), 73–86. See also Marvin Rosenberg, *The Masks of Macbeth* (Los Angeles, 1978), and Dennis Bartholomeusz, *Macbeth and the Players* (Cambridge, 1969).

this play one of the most impressive moral lessons which the stage exhibits.[1]

Kemble himself was less notable than his sister; few leading actors have ever been distinctively identified with this role. Edmund Kean, a Byronic actor with hypnotic eyes and athletic limbs, developed the incipient romanticism, but his Macbeth was far less popular than his Richard III. Kemble's historicism became a showy antiquarianism with Kean's son Charles in the middle of the nineteenth century, when Scottish baronial had become the playground of the rich. The whole production (which he somehow contrived to transport for a command performance at Windsor Castle) was a spectacular 'historical' reconstruction of the kind which Cecil B. de Mille passed on to the cinema. Kean's introduction to his text (1853) is about the accuracy of buildings and costumes in early medieval Scotland which, for lack of records, he was forced to compile from what was known of the Danes and Angles, and from romanticized ruins: on Hecate's entry *'The Scene Dissolves into a Mist'*, and when she exits *'The Mist Disperses, and discovers a Bird's-Eye View of The Island of Iona'*; like de Mille too, Kean deployed hordes of extras and multiple scene shifts (so, of course, did editors intent on specific locations). He did, however, have a serious conscience about textual accuracy, and justified the musical accretions to Act 2 with a lengthy footnote explaining that though it was certainly not Shakespeare's, it was 'an introduction which time and habit have so grafted on the play, that it is very doubtful whether the omission of such a powerful musical effect would not be considered a loss by the general public'. Kean's triple negative betrays his unease at flouting scholarly devotion to textual purity (the later scenes for Hecate had not yet been seriously attacked), but his reason is interesting.

For Kean, *Macbeth* was no longer an 'opera', for opera itself had changed greatly. Verdi produced his *Macbeth*, far the most successful of numerous operatic versions, in the 1840s, a music-drama based on the stage versions current in the London theatre; remote indeed, as reviewers constantly remind us, from Shakespeare, but by no means so remote from stage tradition of

[1] *Macbeth* ... As performed at the Theatres Royal, Covent Garden and Drury Lane. Printed under the authority of the Managers from the Prompt Book. With Remarks by Mrs. Inchbald (1808), pp. 3–4.

10. Design for Hecate and Spirits by George Scharf, Covent Garden, 1837

11. Charles and Mrs Kean in 2.2, Princess's Theatre, London, 1853

45

12. Design for Chorus of Witches by T. Grieve, Princess's Theatre, London, 1853

the previous 200 years. It was seen in London, like Charles Kean's, in the West End; but a third divergent development of major interest was staged in working-class Islington by Samuel Phelps, at Sadler's Wells between 1844 and 1865. The main diet of his theatre was popular melodrama, and his highly successful revivals of Shakespeare were not gentrified. The Porter was at last seen again, and Macduff's son was murdered on stage; the music was severely reduced, but Phelps concentrated on illusion effects, largely with the use of gauzes to make the witches appear and disappear. Phelps's interpretation seems to have been based rather on an understanding of his audience than on subservience to scholarly opinion (of which he was no doubt aware), and his innovations had no effect elsewhere, though they were noticed by serious reviewers. His mixed audience was probably closer in social composition to Shakespeare's than any other since the early seventeenth century.[1] It was the far more academic ideas of William Poel at the end of the century, combined with the penury of Benson's touring company, which pioneered the fashion for simplified productions in this century. Poel's texts were nothing

[1] Henry Morley, *The Journal of a London Playgoer* (1866; 2nd edn. 1891, repr. Leicester, 1974); pp. 137–9 describe Phelps's performances and his audience.

like so 'pure' as he claimed, and the suburban theatres in which he performed bore little resemblance to the thrust stage of the Globe; he did restore the Porter, but not for a popular audience: he played to a coterie of intellectuals and aesthetes. It is symptomatic that it was he, and not Phelps, who primarily motivated the final disappearance of Hecate from twentieth-century revivals. Irving, in the 1880s, had commissioned fresh music, but he retained a chorus; Tree's revival in 1911 was still characteristically spectacular, but since the First World War the play has almost always been given in the severely simplified form assumed for its original production.

The conditions of that original are very difficult to reproduce when the tradition of a daylight theatre has been so long dead, and it is not surprising that, though quite often played, the play has so seldom been successful in modern theatres. Occasional open air performances have been unsympathetically criticized for an unsuitable choice of play. Gielgud and Richardson both failed, for all their skills, to rescue it; Olivier (Stratford, 1955) is remembered by those who saw him, chiefly (it seems) for his resemblance to Edmund Kean in dynamic presence and blazing eyes, together with his own characteristically eccentric locutions. In 1974, Trevor Nunn directed it at Stratford (and later in the Aldwych in London) with the house lights on for much of the play; the experiment puzzled both critics and audiences, and the director offered no explanation in the programme. Two years later he directed it again, this time in the small studio theatre at Stratford, The Other Place, with outstanding success. In this space, the audience is inevitably lit by lights, no matter how they are concentrated on the central playing area, so that, however unconsciously, they accept the illusions in much the same way as had audiences at the Globe. The odd stress on religious rituals which had been tedious in the main theatre was neutralized here because it seemed to be a convenient device to organize the play in what was very nearly a round space; Ian McKellen and Judi Dench gave fine performances, but they were not what made the play so memorable: audiences, even though many must have known the play since suffering it at school, found themselves leaning forward to know what happened next. Samuel Johnson was vindicated: it was the narrative, not the characters of the protagonists, that prevailed, with extraordinary tension.

13. Laurence Olivier and Vivien Leigh at Stratford-upon-Avon, 1955

14. Banquo (John Woodvine) and Macbeth (Ian McKellen) with the Witches in 1.3, directed by Trevor Nunn, The Other Place, Stratford-upon-Avon, 1976

The narrative that emerged does not deal primarily in events as facts, but much more in the forms of illusion that create the facts, above all, that create a tyrant and destroy both him and his victims. At all points 'what is' becomes alarmingly ambiguous, and 'what is not' dominates the event. The supposed restoration in Malcolm's 'triumph' is as perfunctory as the reconciliations at the end of *All's Well That Ends Well*: the thanes may be satisfied with their anglicization as earls, but the audience cannot be. We are left with the image of a severed head: the play's dramatic triumph is to deploy spectacular illusions so as to enforce, not weaken, a tragedy at once personal and political. It has been shown to thrive in studio theatres; in large houses it continues to be less successful. An attempt was made to reintroduce Hecate at the National Theatre in 1967, by Peter Hall and John Russell Brown, but failed because of the belated discovery that the building had not the structural resources to sustain human flight; yet in more adequate theatres there must be a strong case for reviving, not Davenant's heroic adaptation, but those musical developments which he extended from their Jacobean origins, and which proved successful with the Folio text throughout the eighteenth and nineteenth centuries. We need to have more than one version of the play for the same reason as Shakespeare's company did, to serve two very different kinds of theatre.

5. *Text*

There is only one early text of *Macbeth*, in the First Folio of Shakespeare's complete works, which appeared in 1623. It had taken three years to prepare in Jaggard's printing-house in London, and *Macbeth* was probably printed in the second quarter of 1623.[1] It is, in this text, the shortest of all Shakespeare's tragedies, and of any of his plays except *The Comedy of Errors* and *The Tempest*. The text is a good one, overall, which needs very little verbal emendation, and since it is well supplied with stage directions it is usually thought to derive from the theatre prompt-book, probably via a copy omitting any passages that had been cut in performance.

It has, however, two problems: the first is the number of

---

[1] See next note.

lines which are obviously not correctly divided, so that they look like irregular verse; many, but not all, can be quite easily rearranged and do not seem to be otherwise defective. The second is the sequence of scenes from Act 3, Scene 5, to Act 4, Scene 1, inclusive—the part of the play in which Hecate appears—which involves several confusions, where the stage directions are notably inadequate, and which frankly cannot be performed as it stands without at least minor adjustments.

(i) LINEATION

This is more fully discussed in Appendix A where all substantial passages are collated and annotated. The monumental study of Shakespeare's First Folio by Charlton Hinman[1] confirmed that *Macbeth* was set entirely by the two principal compositors who worked on the book, known to bibliographical scholarship as Compositor A and Compositor B. Compositor A set nearly all of the first half of the play to the end of 3.3, and Compositor B all the rest, from 3.4 to the end; he also helped A with two and a quarter pages before his main stint began. It is clear that they divided a single manuscript exactly in half, and each worked on the half he held. When they began *Macbeth* some work remained to be done on *Coriolanus*, and they were able to work simultaneously on both texts; but when that was finished the next tragedy (*Hamlet*, not printed till much later) was apparently not yet prepared for them, and that is no doubt why B helped out where he could simultaneously see A's half of the manuscript, until he was able to carry on with his own.

This is important because the lineation problem is almost entirely confined to the first half of the play, that is, to the part set by Compositor A. Recent studies of his characteristics have concentrated largely on verbal inaccuracy, but some evidence suggests already that he had peculiar habits with lines (see Appendix A). B can be cavalier with words and tends to substitute his own, but he evidently respected verse lines—so much so that he tended (as most later editors have done) to print good prose as if it were bad verse. The only lineation problems in his stint derive, I think, from that. A, on the other hand, seems to have

---

[1] Charlton Hinman, *The Printing and Proof-Reading of the First Folio of Shakespeare*, 2 vols. (Oxford, 1963), i. 358–63.

had no aural sense of verse at all, only a vague visual layout that looks like verse even when it is not.

The Folio was printed in double columns, which meant that fairly often there was not space for a long line, especially at the beginning of a speech where the speaker's name took up extra space (the margin wasn't used). Both compositors treated this problem in the same way: they took the leftover words to the beginning of the next line and used an initial capital as if it were a new verse line. For B it was not: he printed only the leftover words and then started a fresh line quite correctly with the next full verse line. A, on the other hand, filled out his short line with words from the line after and might print as much as ten lines, arbitrarily divided and arhythmic but looking like verse, before he got back on track again.

It is now quite clear that the problem derives solely from the bad habits of Compositor A and not at all from the manuscript which they both used. Correcting the lineation is often quite easy, and editors agree on how it should go; but sometimes it is not so easy, largely because there are anyhow in *Macbeth* a fairly large number of half-lines, or short lines. In a few cases, nearly three centuries of editorial effort has led to no consensus, and these usually occur where, to my ear, the Folio reading is quite satisfactory on the stage; these I allow to stand uncorrected.

(ii) ACT 3, SCENE 5 – ACT 4, SCENE I

There are two principal problems here: the first, and best known, concerns all the material to do with Hecate; the second concerns an apparent contradiction between 3.6, where it is made clear that Macbeth knows of Macduff's flight, and the end of 4.1 where he is told of it and reacts with violent shock. The two are wholly unrelated to each other, but, coming in the same section of the play and linked with an unusually inadequate provision of stage directions, do strongly suggest a different status for the text here. Some passages elsewhere have been suspected of interpolation, notably the episode of the King's Evil, 4.3.140–59, chiefly on the grounds that the text will be perfectly coherent if it is omitted; but, as Muir pointed out,[1] its dramatic function is evident—and the text we have is itself coherent (l. 140 is short, but creates no

---

[1] Commentary note on 4.3.140–59 (Introduction, p. xxxii, refers only to ll. 140–6).

problem on stage). What distinguishes 3.6 and 4.1 is that they cannot be performed as they stand, and so with 3.5 form a unit that differs from any other part of the play. It lies entirely within the part set by Compositor B, but the untidiness does not correspond with any of his known habits, so it must be attributed to the manuscript he was using. It seems as though the material available for these three scenes consisted of working sheets of a revision not finally tidied up for performance.

This could quite easily be the result of a late decision to alter this part of the play, chiefly no doubt to introduce Hecate, but not exclusively for that. Act 3, Scene 6, is a brilliantly written scene for Lennox and 'another Lord' in which they have to fence with words because they dare not trust each other. It is perfectly coherent until l. 37, and thereafter it becomes confusing. F reads:

> And this report
> Hath so exasperate their King, that hee
> Prepares for some attempt of Warre.   (37–9)

That must mean the King of England, who has been described in ll. 27–37. But in l. 40 Lennox asks 'Sent he to Macduff?', where 'he' must be Macbeth, though he has not been mentioned since l. 25. If the report is to the English King, then Macduff must already have arrived, but in ll. 46–8 Lennox is praying for an angel to fly to England to accelerate Macduff's message.

To solve this Hanmer emended 'their King' to 'the King', intending this to refer to Macbeth, not Edward the Confessor, and most editors have followed him. But the only king under discussion has been Edward, and an audience can hardly be made to understand the switch. Furthermore, in Act 4, it is Edward, not Macbeth, who is preparing for war. Lastly, this change makes it decisive that Macbeth knows of Macduff's flight, which in 4.1.155–70 he certainly does not—and it is not credible that his shock there should be simulated.

In the theatre 3.6 has often been omitted altogether (following Davenant's adaptation), or a cut version of it tacked on to the end of 4.1. It has, indeed, been argued that this was its original position, from which it was shifted back to divide 3.5 from 4.1; but it would always have been needed to separate 3.4 from 4.1. Muir argues that it deals with events which have not yet taken place, but that is not clear.[1] In Act 4, as in many of Shakespeare's plays, time becomes much more diffuse after the concentration of Acts 1 to 3, and 3.6 serves as a chorus to divide the play, and to offer hints of the eventual retribution; placed here the scene serves a very similar function to 3.1 in *Lear*. In *Lear*, Kent and a gentleman talk of Cordelia's landing in Dover just before Lear descends into total madness and Gloucester is blinded;[2] in *Macbeth*, Lennox and another Lord talk of English invasion and Scottish defection just before Macbeth has his final encounter with the Weïrd Sisters and resolves on open violence with the murder of Macduff's family. It would be better, I think, to adjust it where it stands, possibly by cutting ll. 37–9, though that would still leave some confusion.

Hecate is a different matter. Her presence, and her utterance, have been largely disapproved of by editors although her scenes, somewhat expanded, formed the core of the operatic development of the play which held the stage until this century (see section 4 above). She is a goddess from a machine, and as such has a different verse from the Weïrd Sisters; exception has been taken on that account alone, but Jupiter in *Cymbeline*, and Ceres and her colleagues in *The Tempest*, equally use a different verse form from others in their plays. It is hardly a serious objection that she adds nothing to the plot, for her function is rather spectacular than narrative, and it seems that she presides over the climactic apparition scene, which is appropriate. Her reproach to the Weïrd Sisters in 3.5 may be slightly obscure, but it serves to establish her as the initiator of the apparitions in 4.1. Act 3, Scene 5, is strikingly well written; there is no good cause to question its right to be there.

Hecate's entry into 4.1 is less satisfactory: the stage direction gives no hint how her entry should be effected, nor does it identify the 'other three Witches'; the chorus quickly became much larger than three, and may have been so from the start if 'three' here is an accidental repetition of the other entries for witches. Whether that is so or not, they are presumably very different from

---

[1] Introduction, p. xxxiv.
[2] *The History of King Lear* (1608), scene 8, ll. 21–4. This scene was also revised in *The Tragedy of King Lear* (1623), 3.1, where the French landing at Dover is displaced by discussion of domestic spying, leaving the allusion to Cordelia somewhat obscure (both versions are in Oxford).

the Weïrd Sisters, and (especially if they were juveniles like the 'little spirit' of 3.5) may therefore justify the apparent incongruity of line 42 'Like elves and fairies in a ring'. But there are other anomalies: immediately before and after this entry, speeches are allocated to the Second Witch who elsewhere only speaks in turn after the First Witch who always leads; Hecate is never directed to exit unless she is included in the general direction to dance and vanish (how?) 100 lines later, during all of which she has not spoken a word; lastly, the final speech of the First Witch (ll. 140–7) is quite alien to her usual tone but precisely in Hecate's style. It may be that some renumbering of witches took place, because Hecate once on stage becomes properly '1 Witch', and the First Witch becomes '2 Witch' accordingly; this would account for the unlikely prominence of the Second Witch in ll. 37–61, and for the attribution of what is surely Hecate's final speech to the First Witch. But such renumbering has certainly not been carried into the intervening apparition scene, where the sequence 1, 2, 3 is carried on in the usual way and there is no new '4 Witch'.[1]

It does seem clear that Hecate's appearance in 4.1 is the result of a revision of a scene that originally was written without her. But she cannot simply be deleted, because evidently (as with 3.6) some original material has been excised to fit her in: the Second Witch would still speak out of turn, and there would be no response to Macbeth's final appeal. And if she is an intrusion in 4.1, then it follows that 3.5 is also added material; it can simply be omitted—as it usually is on the twentieth-century stage.

The songs in 3.5 and 4.1 are indicated in the Folio only by their first lines, another anomaly since all other songs in the plays are given in full. It is even more strange that texts for both of them can be found in a manuscript copy of a play of Middleton's, *The Witch*;[2] it has a number of echoes of *Macbeth* in it, and its scenes of bawdy witchcraft seem to burlesque Shakespeare. Almost identical texts of the songs appeared in Davenant's adaptation of *Macbeth*, printed in 1674, so they were clearly associated with it

---

[1] See E. B. Lyle, 'The Speech-Heading "1" in Act IV Scene 1, of the Folio Text of *Macbeth*', *The Library*, 25 (1970), 150–1. Lyle proposed that all the First Witch's speeches in ll. 76–90 should also be Hecate's, but they read as if written before Hecate was introduced.

[2] Bodleian Library, Oxford, Malone MS 12; ed. W. W. Greg and F. P. Wilson, Malone Society Reprints (Oxford, 1948–50).

throughout the seventeenth century. The song in 3.5 survives also in two collections of manuscript lute songs from the same period as the manuscript of *The Witch*, and in a quarto of *Macbeth* printed in 1673. The complex problems resulting are discussed, and the music given, in Appendix B.

There are too many unknowns about these scenes to allow a confident guess about their origin: when Hecate was introduced, when Middleton wrote *The Witch*, why 3.6 is in a state of imperfect revision though Shakespeare's responsibility for it is not questioned, why the songs were not printed in full. The questions of dating and authorship are discussed in sections 6 and 7 below. The simplest explanation I can offer is that Shakespeare and Middleton collaborated on the revision, Middleton contributing the songs and possibly Hecate's speeches, and that a tidy copy was never made before their work was incorporated into the prompt-book in the theatre; in the first instance the song sheets may have been in the hands of the musicians, but at some point they must have been included with the rest, or Davenant would not have found them; it remains odd that they were not available to the Folio printers. The assumption here is that Shakespeare accepted the Hecate material even if he did not write it, and that seems to me extremely probable.

(iii) THE BREVITY OF *MACBETH*

The unusually short text has been attributed to two causes: first, that it derives from a version cut for performance (there is no evidence whatever for the common assertion that it was for a performance at Court, which is most unlikely; see pp. 72–6); second, that as compression is a feature of its language, so it is also of its whole structure.

The second does not impress me: it is certainly true that the language of *Macbeth* is distinctive, and that it can be extraordinarily densely wrought as well as elliptical, especially in the very complex soliloquies; but it is also true that a superficially similar language is used for merely prolix courtesies in several court scenes, and in such dialogues as Malcolm's with Macduff in 4.3. The same can be said of much longer plays, especially late ones.

On the other hand it does, as I have said, appear to derive from a copy in which theatrical cuts have been respected (then, as now, it was usual for printers to ignore playhouse cuts and to give as many as possible of the author's words). It is also possible that the censor (the Master of the Revels) demanded some cuts or alterations, since the play touches politically sensitive matters. None of the missing material, even whole scenes, that has been guessed for it is convincing, nor should it be, for it is the business of theatrical cutting to leave out nothing of narrative importance, and to end up with a perfectly coherent text for performance. If the 'bad quarto' of *Hamlet*, which certainly derives from performance, were fleshed out with more perfectly remembered words it would be only slightly longer than *Macbeth* and few would notice the omissions. I take *Macbeth*'s brevity to be entirely a matter of its derivation from performance.

My conclusion is, then, that the play as we have it is a generally sound, though not perfect, text of what was performed between about 1610 and 1620; behind it we can see the lineaments of an earlier version in which Hecate did not appear. We cannot revert to that (because some of it has been lost or altered), even if we wished to.

I do not think we should wish to: the old belief that editors should act as archaeologists and disinter what the author 'originally intended' was a heresy, for a play does not have finally determined intentions before it has reached the stage, and changes are made again with revivals; that was as true then as it is now: there are at least two versions of Beckett's *Waiting for Godot* in print, and at least three of Miller's *The Crucible*—and in neither play can rehearsal proceed if actors hold different editions. Some of Shakespeare's plays, where two early editions survive, have the same characteristic: *Richard III* and *Hamlet* for instance, and for *King Lear* the new Oxford *Complete Works* gives two versions of the play entire. What we do need to understand is the function of the text we have: I take *Macbeth* to be a fairly reliable record of what the play was at one point in its early history. It is not possible for an editor to go beyond that, for the choices to be made depend on the company performing it: they will be forced to make some adjustments, whether they decide to have Hecate in or to leave her out; no doubt any production will involve more cutting and adapting than just that, to suit the casting, the theatre, and above all the very different society in which they must perform it; and it is likely that in the process fresh discoveries will be made.

6. *Middleton and the Revision of* Macbeth

At least since the Clarendon edition of 1869, it has been frequently suggested that Middleton was responsible for 3.5 and the Hecate passages of 4.1. Middleton's hand was also suspected elsewhere in the play, wherever editors from Clarendon to Wilson felt the writing to be 'certainly inferior' (Clarendon) or the work of a 'botcher' (Wilson). Middleton was a highly professional dramatist, capable like any other (including Shakespeare) of inferior work, but his best plays are among the finest in a period of great drama; whatever reasons there may be for associating Middleton with the play can have nothing to do with inferiority as such. His name was canvassed originally because of the songs in *The Witch*, and he acknowledged that as his in a prefatory note to the manuscript copy which survives.

There has been a revival of interest in authorship studies after long disfavour for the fashion of 'disintegration' which spilt so much ink in the late nineteenth and early twentieth centuries. The text is decidedly more important than the identity of the author(s), but, deplorably, attribution still induces critical prejudice. New work has been stimulated by the possibility of more comprehensive and more sophisticated use of statistical evidence with computers. Gary Taylor, in the *Textual Companion* to the Oxford *Complete Works* (1987), reviews carefully what is so far available and finds only one test of particular value for determining authenticity. So-called 'function words' can be used with sufficient idiosyncrasy in English to identify the user; Shakespeare apparently shows sufficient consistency in the use of *but, by, for, not, so, that, the, to, with*. Substantial deviance from his norms can indicate the presence of another hand, though it cannot identify that hand. Taylor used a minimal core canon for comparison, excluding any plays at all widely suspected of collaboration; from this core he excluded *Macbeth*. Obviously its inclusion would affect the statistical results, but only by a very small percentage, since his core consisted of thirty-one plays. Passages as short as the Hecate material are not sufficient to establish statistical probability, but setting *Macbeth* without Hecate against the core canon he found that the figures 'arouse some anxiety': 'the high deviance for the word *by* suggests that the remainder of the play may also have been affected verbally by adaptation'.

This view is supported by Dr R. V. Holdsworth in a forthcoming book on Middleton's collaboration with Shakespeare. He has, with generous hospitality, given several hours to outlining his material to me. This consists principally of echoes of Middleton's idiosyncratic habits, both with stage directions, and with language. For the first, Holdsworth has carried out exhaustive research into the occurrence of stage directions in the precise form 'Enter A meeting B'. Most dramatists, including Shakespeare, tended to avoid this formula, perhaps because it leaves unclear whether 'B' is also entering or is already on stage, preferring such alternatives as 'Enter A and B severally' or 'Enter A at one door, B at the other'. Middleton's twenty-two accepted plays and masques have the formula ten times, including two examples in the autograph manuscript of *A Game at Chess*. In all ten cases 'B' is meant to be entering as well. In the other 634 plays in the period 1580–1642, including masques and shows which employ theatrical directions, there are only twenty-seven instances of the formula; many are in plays by Heywood, and ten of them assume that 'B' is already on stage. Of the remaining seventeen, three are in plays by Shakespeare: one in *Timon* (in a scene Holdsworth assigns to Middleton), and the other two in *Macbeth*, at the openings of 1.2 and 3.5. In the text, Holdsworth has collected a very large number of phrases and of verbal idiosyncrasies paralleled in Middleton's work; few, as he says, would be convincing by themselves, but they might be significant cumulatively. There seems to be a particular concentration in two speeches: the Bleeding Captain's in 1.2.7–42, and Macbeth's soliloquy after the apparitions in 4.1.159–70; otherwise the distribution is random and fragmentary, and Holdsworth finds no clear pattern of collaboration such as he believes happened between Shakespeare and Middleton in *Timon*. Act 1, Scene 2, has often been suspected in the past, but on unsatisfactory grounds: the irregularities in the Captain/Sergeant's speech are simply the result of Compositor A's illiteracy, and its stylistic difference has a clear dramatic function in linking the rhetoric of the classical messenger to a badly wounded speaker; Macbeth's soliloquy is not usually doubted, but it is of course possible that the revision of 4.1 required some adjustment to its ending.

The songs remain the one decisive link with Middleton. Holdsworth finds that he habitually reused his own material, very seldom other people's. In *More Dissemblers besides Women* (1619) he reused two songs from his own earlier plays, *A Chaste Maid in Cheapside* (1613) and *The Widow* (*c*.1616). If Holdsworth's other evidence gains general credence, it will support the belief that Middleton was responsible for the revision of *Macbeth*, probably touching the play at other points while incorporating his new material; but it remains mysterious that his hand should show in so many places in a play for which the general responsibility was, unquestionably, Shakespeare's.

7. Dates

(i) THE ORIGINAL PLAY

Since Malone at the end of the eighteenth century, *Macbeth* has usually been dated 1606, and the vast majority of critics has found that this conforms with a sense of its relation to Shakespeare's other plays, soon after *Lear* and close to *Antony and Cleopatra*. Occasional attempts to argue for an earlier version have not proved convincing. There is no evidence to contradict 1606, but there is also very little to support it. The fact that interest in Scotland was intensified by James I's accession in 1603 does not necessarily point to a date after that, for his succession had been canvassed well before and his mother, Mary, had been a notoriously controversial figure both before and after her execution in 1587. Holinshed's *Chronicle of Scotland* was first published in 1577 and reprinted (in the edition Shakespeare used) in 1587. The story of Macbeth is in that, and was presumably the subject of a ballad of Macdobeth, now lost, referred to in the *Stationers' Register* on 27 August 1596, and by Kemp in *Nine Days' Wonder* in 1600: 'I met a proper upright youth ... a penny Poet, whose first making was the miserable stolen story of Macdoel, or Macdobeth, or Macsomewhat, for I am sure a Mac it was, though I never had the maw to see it' (ed. Dyce (1840), p. 21).

There is no reason, therefore, to see *Macbeth* as particularly related to James's accession in 1603, but three slight allusions do bring it closer to 1606. The Jesuit Superior in England, Father Henry Garnet, was tried in March 1606 for treason because of his involvement in the gunpowder plot to blow up the King in Parliament on 5 November 1605. Garnet was proved to have committed extensive perjury in his evidence, and justified himself on the grounds of a right to equivocate in self-defence. He was hanged on 3 May 1606, and the breaking of oaths made on the Bible became a common topic of political debate to which the King himself contributed. In the eighteenth century Warburton pointed out the connection between this and the Porter's obscure allusion in 2.3.8–11: 'Faith, here's an equivocator, that could swear in both the scales against either scale, who committed treason enough for God's sake, yet could not equivocate to Heaven: O come in, equivocator.' This passage certainly looks like an allusion. Garnet was said to console himself in prison with wine, and accused of fornication with his friend Mrs Vaux, and that may contribute to ll. 28–34: 'Therefore much drink may be said to be an equivocator with lechery: it makes him, and it mars him . . . in conclusion, equivocates him in a sleep, and giving him the lie, leaves him.' Shakespeare very rarely used the word 'equivocate' elsewhere, though he does repeat it once in this play; in 5.5.42–4 Macbeth responds to the 'moving' of Birnam Wood:

> I pull in resolution, and begin
> To doubt th'equivocation of the fiend
> That lies like truth.

There is nothing remarkable in the use of the word here; it is, if anything, more surprising that it isn't used more often, for equivocation can properly be seen as a fundamental preoccupation in the play. Macbeth equivocates with his conscience from his first appearance in 1.3; Lady Macbeth with her humanity in 1.5 and after; the Weïrd Sisters in all their predictions; Malcolm, Donalbain, and all the thanes, including Banquo, after the murder of Duncan; Malcolm with Macduff in 4.3; and so on. It is certainly conceivable that the theological and political scandal of Garnet's trial contributed to this leitmotif in the play, though only the Porter's allusion is directly relevant to dating. It cannot have been made before May 1606, though of course it could have followed at any time after that—the issue was remembered for years.

A second allusion is more tenuous. Holinshed lists a number of Malcolm's acts of gratitude to the thanes who supported his rebellion against Macbeth; Shakespeare selected only one of them:

> My thanes and kinsmen,
> Henceforth be earls, the first that ever Scotland
> In such an honour named.   (5.7.92–4)

'Earl' was an English title; James's profuse conferring of English titles on his Scottish followers caused frequent satiric comment, however necessary it may have been to his ambition to unite the monarchies. Chapman referred to it in *Bussy d'Ambois* (c.1604), 1.2.154, and in 1605 was imprisoned with Marston and Jonson for a similar allusion in their *Eastward Ho!* Shakespeare's allusion, if it is one (it does not seem to have been discussed before), is more circumspect and, being historical, less likely to raise objection; but it probably springs from the same circumstances. If so, it sorts oddly with the common contention that the play was intended as a compliment to James. This is fully discussed below, pp. 71–6; though the play certainly dramatizes the Scottish monarchy, and shows the Stuart line, it does not do so in a light best calculated to flatter James. It is most unlikely that it was ever designed for court performance, and there is no evidence whatever that it was given before the King of Denmark on 7 August 1606. Indeed a third, much more specific, allusion makes it very unlikely that it could have been.

In 1.3.4–25 the Weïrd Sisters plan all the revenge they can on the sailor's wife who had abused them: 'Her husband's to Aleppo gone, master o'th' *Tiger*' (1.3.7). Hakluyt[1] and his successor Purchas[2] give details of numerous voyages by ships called *Tiger*, of which there must have been at least three between 1564 and 1606, to judge from occasional references to tonnage: one of 50 tons, one of 140, and one of 600. There were probably more; in any case it was clearly a popular name for ships, and the reference might be casual; but two voyages do seem to have contributed. The first, in 1583, carried an ambitious trading expedition to Tripolis on the coast of Syria; from there John Eldred and others proceeded overland to Baghdad; they were dismayed to find that they could do nothing there without money, and returned from Tripolis in the *Hercules* in 1588, after five profitless years. Eldred, John Newberry, and Ralph Fitch all sent letters

---

[1] *The Principal Navigations Voyages Traffiques and Discoveries of the English Nation*, 1598–1600, 12 vols. (Glasgow, 1903).

[2] *Hakluytus Posthumus or Purchas his Pilgrimes*, 1625, 20 vols. (Glasgow, 1905).

home from Aleppo at various dates in May 1583 by the purser of
the *Tiger* who had presumably accompanied them this far. Aleppo
was not, and is not, on the coast, but it is easy to see how the
error arose from a casual memory of the accounts in Hakluyt[1]
(the extent of documentation makes this voyage conspicuous).
Another, and much later, voyage by a *Tiger* is elaborately recorded in Purchas.[2] It set out for the far East from Cowes in the
Isle of Wight on 5 December 1604, and after a voyage fraught
with incidents of storm and piracy (their own as much as anyone
else's), some of them disastrous, it did finally reach Japan. At one
time a smaller ship, the *Tiger's Whelp*, disappeared in a tempest,
but it survived and eventually rejoined the fleet; it may have
contributed to the significant reference to the limited powers of
the Sisters:

> Though his bark cannot be lost,
> Yet it shall be tempest-tossed.   (24–5)

The remains of the fleet finally entered Milford Haven in south-
west Wales on the 27 June 1606. E. A. Loomis has pointed out
that if the days of entering and leaving harbour are discounted,
they were absent for 567 days which equals $7 \times 9 \times 9$, exactly as
in the Sisters' magical chant:[3]

> Weary sev'n-nights, nine times nine,
> Shall he dwindle, peak, and pine.   (22–3)

The coincidence is surprising, and hard to ignore. If it is
significant, then this passage cannot have been written before the
beginning of July 1606 (time has to be allowed for the news to
reach London). If Shakespeare wrote sequentially (which we do
not know), he cannot have begun the play much sooner. Topical
allusions like this, or the Porter's references to Garnet, can always
be added to a play; but neither looks like an addition, and both
have essential dramatic relevance.

This suggests composition rather later in 1606 than has
usually been assumed; but if the imaginary court performance in
August did not take place, this is perfectly possible. There is no
satisfactory evidence to limit the latest date by which *Macbeth*
must have been on the stage, before Forman saw it at the Globe in

---

[1] Op. cit. vi. 1–9, v. 455, 465.   [2] Op. cit. ii. 347–66.
[3] E. A. Loomis, 'The Master of the Tiger', *Shakespeare Quarterly*, 7 (1956), 457.

1610 or 1611 (see Appendix C). Echoes have been suggested in three plays of 1607, but none of them will survive scrutiny. *The Puritan*, now confidently attributed to Middleton,[1] was entered in the Stationers' Register on 6 August 1607; at the end of Act 4 a dead corporal rises, and Sir Godfrey comments to the Sheriff 'instead of a jester, we'll ha' the ghost i'th' white sheet sit at the upper end a'th' table'. This never did strongly suggest Banquo's ghost, and now Holdsworth has pointed out that Middleton had much earlier had a ghost 'Sit ... at the upper end of the table' in his satire *The Black Book*, printed in 1604. *Lingua*, an anonymous allegorical comedy, was entered in the Stationers' Register on 23 February 1607 (1606 old style); it has a sleep-walking scene, but it bears no close resemblance to Lady Macbeth's, and the discussion that follows is based on Scaliger, which the Doctor's disquisition in *Macbeth* is not. Beaumont's *The Knight of the Burning Pestle* has, in 5.1.18–28, a passage which is rather more like Macbeth's encounter with Banquo's ghost, but the play was not printed until 1613 and its original performance is unknown; Muir said 'probably acted in 1607' (p. xvii), but M. Hattaway in the New Mermaid edition (1969) prefers 1608 while stressing that there is no certainty.

I have argued in this Introduction that *Macbeth* was written for the Globe Theatre, before the move into the Blackfriars was contemplated, and if that is correct, the latest possible date would be 1608; earlier would be more probable. It is generally assumed to have been written before *Antony and Cleopatra* which was entered in the Stationers' Register 20 May 1608. Barnabe Barnes's tragedy *The Devil's Charter*, performed in February 1607, makes an elaborate allusion to Cleopatra's death, complete with asps; but there is no verbal echo, and the story was well-known. If he was alluding to Shakespeare, then it would appear that *Macbeth* and *Antony and Cleopatra* were both written in the second half of 1606. Given Shakespeare's fertility this is possible, but even for him it is hardly probable. The only evidence available to determine priority between the two plays is metrical and verbal tests which cannot determine priority between successive plays which are in subject and style in such extreme contrast. Two allusions in *Macbeth* (3.1.56, 5.7.31)

[1] D. J. Lake, *The Canon of Thomas Middleton's Plays* (1975), 109–35; composition is confidently dated 1606.

may refer to *Antony*, but may simply depend on recollection of Plutarch.

My conclusion can only be tentative: that *Macbeth* was probably written in the second half of 1606; close, no doubt, to *Antony and Cleopatra*, but whether before or after I do not know.

(ii) THE REVISED VERSION

The date at which Hecate and her song and dance team were introduced depends entirely on the relation between *Macbeth* and *The Witch*, since the songs appear in both plays. The manuscript of *The Witch* is a copy made for sale by Ralph Crane who was frequently employed on play scripts from 1619 onwards through the 1620s. The title-page describes it as 'long since acted, by his Majesty's Servants at the Blackfriars'. Middleton added a dedication in his own hand, so the information is fairly trustworthy; 'long since' implies a date before 1620, and the King's Men did not use the Blackfriars until 1609. Middleton described the play as 'this (ignorantly ill-fated) labour of mine', which used to be taken as meaning a theatrical flop; but Middleton went on 'Witches are (ipso facto) by the law condemned, and that only (I think) hath made her lie so long, in an impassioned obscurity'. This clearly attributes its withdrawal, not to unpopularity, but to some intrusion of law after it had been allowed for performance by the censor, possibly a complaint from the court. The complex tragi-comic plot involves the impotence of an unwanted husband, procured by witchcraft, and used as grounds for 'divorce' (annulment). The motif was used by Tourneur in *The Atheist's Tragedy*, printed in 1611; but in Middleton's play it is so developed that it most probably alluded to the notorious divorce proceedings between Frances Howard and her first husband, the Earl of Essex, in which James was much involved. She claimed that Essex was impotent with her and not with other women, which Middleton's plot echoes, but Tourneur's does not. When Frances and her second husband, the Earl of Southampton, were tried in 1615 for the murder of Sir Thomas Overbury (in connection with the divorce proceedings), she was convicted of conspiring with Simon Forman to procure witchcraft for her husband's impotence. Middleton ascribes the witchcraft to a third party, so A. A. Bromham argued that he probably wrote before the full details emerged in 1615.[1] Anne Lancashire pleads persuasively for a fuller critical study of his play as a whole, where the plot is not carried through and the parties are reconciled; she suggests that the purpose was reformatory, in which case the play should be dated before the divorce was granted in 1613. She cites a letter written in 1610 which makes it plain that rumours of Frances's intent to poison Essex were already rife then, but I find her further claim that Tourneur imitated Middleton rather than the other way round less convincing.[2] On this showing *The Witch* might well have been written in or before 1613; but it is hardly conclusive that it was earlier than 1615, when Middleton began to write regularly for the King's Men.

One piece of evidence that points in this direction does not seem to have been considered in previous accounts. Settings for two of the songs in *The Witch* survive: one is probably by Robert Johnson, 'Come away, Hecate', discussed in Appendix B; the other is by John Wilson and is not used in *Macbeth*. Johnson composed for the King's Men from 1609 to about 1615; Wilson from 1615 or slightly later. Wilson was born in 1595 so, however precocious he may have been, it is unlikely he would have been employed by the company before 1615, and almost certain he would not have been before 1613.

The question of which was done first, *The Witch* or the Hecate material in *Macbeth*, has been much argued, but recent opinion has generally given precedence to *The Witch*, chiefly on two grounds. First, that since *The Witch* failed, Middleton was ready to take material from it for revising *Macbeth*; second, that the material is quite alien to *Macbeth*, whereas the singers in *The Witch* have roles in the play, and the matter of the songs relates to the plot. The first assumes that Middleton would not have reused songs from a successful play, which in fact he did (see pp. 58–9); the second, it seems to me, goes either way. Material for a masque would not normally have much plot connection with the play in which it is performed; and if Middleton transferred the songs, he certainly did not do the same

---

[1] 'The Date of *The Witch* and the Essex Divorce Case', *Notes and Queries*, 225 (1980), 149–52.

[2] '*The Witch*, Stage Flop or Political Mistake?' in '*Accompaninge the Players*': *Essays Celebrating Thomas Middleton, 1580–1980*, ed. Karl Friedenreich (New York, 1983), 161–81.

with Hecate: in *Macbeth* she is an imperious goddess, while in *The Witch* she is the randy leader of a troop. If, on the other hand, the songs had been first in *Macbeth*, they would have provided Middleton with a cast and themes to develop in his entertainingly bawdy burlesque of witchcraft in his play. My feeling is that this is the more probable sequence, and it is supported by Wilson's participation in *The Witch* along with Johnson. The second song may possibly be based on material in Scot's *Discovery of Witchcraft* (1584) which Shakespeare did not use and Middleton did (see Appendix B), but since it is in any case an addition, this does not affect the present argument. If I am right in thinking that the Hecate material was added after the shift to the Blackfriars Theatre, then it was probably done fairly early; Forman saw it at the Globe in 1610 or 1611, so it was probably adapted for the Blackfriars too by then and the same version used in both theatres. Forman does not mention Hecate, but he does not mention the apparitions or their prophecies either, and they must surely have been in the original play.

My conclusion here is even more tentative than about the original play: I suggest that *Macbeth* was revised in 1609–10, and *The Witch* written in 1615.

8. *Sources*

The term 'source' has no precise definition. It is commonly used for the literary materials providing the main matter on which Shakespeare based his play; but other materials which seem to have contributed interestingly are often added, in this case passages from Seneca (section iii below), and I have added discussion of relevant medieval plays in section v. Section ii, on Stuart politics, is not concerned with literary materials, but since they have been seen as the 'only begetter' of the play, they must be treated as sources; in any case the matter follows naturally after the handling of historical narratives discussed in section i. The identification of literary sources depends on clear use of narrative, or on verbal echoes, or both; its critical interest is that it can offer perceptions about the recasting of materials into the final play.

For sections i and iii, copious materials and full discussion are provided in Geoffrey Bullough's *Narrative and Dramatic Sources of Shakespeare*, vii (1973); they are more concisely summarized in

Kenneth Muir's *The Sources of Shakespeare's Plays* (1977). There
have been briefer selections from Holinshed's *Chronicles* in the
Everyman series, and in *Shakespeare's Holinshed*, ed. R. Hosley
(New York, 1968). The standard work on Elizabethan and
Jacobean attitudes to witchcraft (section iv) is Keith Thomas's
Religion and the Decline of Magic (1971).

(i) CHRONICLES

As for his other plays on British history, Shakespeare used
Holinshed's *Chronicles*, of England and of Scotland, in the second
edition which appeared in 1587. The first edition, 1577, was in
two volumes with woodcut illustrations; they were conflated into
a single volume without illustrations in the reprint, and a third
volume (as long as the other two together) was added. There is no
reason to suppose Shakespeare knew the illustrations to the
section on Macbeth, though one is often reproduced, showing
Macbeth and Banquo meeting the Weïrd Sisters: the men are on
horseback, in Elizabethan gentleman's costume, and the Sisters
are well-dressed ladies. Holinshed himself said[1] that his narrative
was a translation of Hector Boethius's Latin *Chronicle of Scotland*
(*c.*1527), using a translation into Scots by Bellenden, printed
about 1536.[2] In fact it is a free rendering rather than simply a
translation into English; on a few details, Shakespeare seems
even closer to Bellenden than to Holinshed, but he could easily
have developed these for himself. Boethius and Holinshed were
conscientious chroniclers, but rarely critical of their material,
much of which was mythical; Buchanan, however, in his *Rerum
Scoticarum Historia* (1582), was sceptical of much that depended
on premonitory dreams or supposed witchcraft. Bullough and
Muir both think Shakespeare had consulted Buchanan, chiefly
because of his comments on Macbeth's mind, though they admit
that the ideas could readily have been developed from Holinshed.
There is even less reason to believe that Shakespeare consulted
either John Leslie's *De Origine Scotorum* (printed in Rome in 1578)

---

[1] *Description of Scotland* (1585–7), title-page and dedication.
[2] Ed. E. B. Batho and H. W. Husbands, Scottish Text Society (Edinburgh, 1941).
'Boethius' was a Latinization of the Scots name Boyis; he also used the half-way
form 'Boece'. Born about 1465, he spent some years at the University of Paris,
studying under (amongst others) Erasmus; they corresponded after Boethius's
return to Scotland. He advised the Bishop of Aberdeen on the foundation of a
university college there, and became its first principal in 1505. He died in 1536.

or William Stewart's *The Buik of the Chronicles of Scotland* (not printed till 1858; James I probably possessed it in manuscript). Leslie illustrated the Stuart line with a family tree leading up to James as prince, since his mother was still alive; but as a source for the show of kings in 4.1 this is by no means as clear as Holinshed's allusions (not in Bullough) which are discussed below. For some plays, Shakespeare seems to have engaged in research, but I doubt whether he did for *Macbeth*.[1]

Holinshed tried to achieve a clear account of very confusing events, assuming a central significance for 'kings' in a society which, despite his efforts, sounds more like a constant warring of rival chieftains only occasionally accepting subservience to one more successful than the rest. The supremacy of the individual had about as much stability as that of a modern prime minister heading a coalition based on proportional representation. Holinshed records efforts to establish lineage by legislation, and obviously sees them as attempts to stabilize government in the manner of the Tudor, or Stuart, dynasties. It is precisely that which accounts for the prominence in the chronicles of the later part of Macbeth's career, for it formed the starting-point of a mythical genealogy which the Stuarts invented for themselves when they achieved monarchy in the fourteenth century. Banquo (who had no historical existence) was said to have been murdered by Macbeth (who had), but his son Fleance escaped to Wales where he seduced a princess; their son, Walter, later fled back to Scotland, where his military success eventually won him the post of Royal Steward, the senior court official. Walter, whose post became hereditary, was the historical founder of the family. Hence their name, which Holinshed spells 'Steward' throughout.

The first part of Macbeth's career is the usual story of wars internal and external; Macbeth was a successful warrior, acting on behalf of Duncan, a relatively young king who is alternately praised for loving peace and criticized for feeble inactivity. Shakespeare adjusts that by giving Duncan the reverence due to old age, and omitting explicit criticism. But he does not simply idealize: Duncan's dependence on generals is not explained, but I have already remarked on the irony of his successive appoint-

---

[1] For Boethius, Buchanan, Leslie, and Stewart, see Bullough, pp. 436–42.

ments of Thanes of Cawdor. Holinshed recounts the first encounter with 'the weird sisters, that is (as ye would say) the goddesses of destiny, or else some nymphs or feiries, indued with knowledge of prophecy' (Bullough, p. 495). Their prophecy is as essential to the myth of Banquo's descendants as it is to the plot of the play. Macbeth did not encounter them again, but learnt of the danger of Macduff either from his over-great power, or else possibly from 'certain wizards, in whose words he put great confidence' (p. 500), because the earlier prophecies had been fulfilled. In other words, he had come to rely on soothsayers, and the remaining predictions came from 'a certain witch, whom he had in great trust' (p. 500). The conflation of all these figures in the Weïrd Sisters is an obvious dramatic concentration, yielding the distinctive ambiguity of their nature, as the goddesses of destiny and as village witches.

Macbeth's wife, hearing of the initial prophecy, 'lay sore upon him to attempt the thing, as she that was very ambitious, burning in unquenchable desire to bear the name of a queen' (p. 496). Boethius was slightly more detailed, adding that she 'called him oft times feeble coward' (p. 496 n. 5), but Shakespeare obviously developed his Lady from Holinshed's account of events a century earlier. Donwald, Captain of Forres Castle, exposed the use of witchcraft to make King Duff fatally ill. Duff duly recovered, suppressed the inevitable rebellion, and executed his prisoners at Forres. Donwald pleaded in vain for some of his kinsmen, but retained Duff's confidence. Donwald's ambitious wife stimulated his resentment and, when Duff paid a friendly visit, they made his two chamberlains drunk, Donwald slit the King's throat, pretended horror when it was discovered, and killed the chamberlains. His severity in pursuit of suspects made some of the lords suspicious, but they did not dare challenge his power. For six months sun and moon never shone, monstrous sights were seen, and horses ate each other.

Duff's death evidently attracted an accretion of mythological detail which Shakespeare transferred to Duncan's, allowed only one sentence in Holinshed. The first part of the account of Macbeth ends with the excellence of Macbeth's rule during his first ten years as king, attributing to him strong rule, and important legislation notably enhancing the central power, and also the status of women by giving limited rights of inheritance to daughters and widows. The selection of items for praise is obviously sympathetic to Elizabethan government.

The quarrel with Macduff involves a complicated story about the building of Forres castle which Shakespeare reduced to refusal of an invitation (command) to dinner, but the deterioration of Macbeth's rule is attributed entirely to Banquo's murder. Holinshed uses that crime to introduce a full account of the entire line of the Stuarts, mythical and historical alike, right through to Mary and James (not yet king). Then he switches abruptly back: 'But to return unto Macbeth, in continuing the history, and to begin where I left, ye shall understand that after the contrived slaughter of Banquo, nothing prospered with the foresaid Macbeth' (p. 499). It seems that Macbeth's later taste for evil depends on the Stuart myth rather than the given facts.

Shakespeare suppressed the good reign, but very possibly transferred the idea of it to the potential in Macbeth's character. It is not quite true, however, that he entirely contracted the years of Macbeth's reign. They are not directly specified, but in the course of Act 4 the time-scale of the play changes radically: Acts 1-3 are continuous, more or less, and the rapid sequence of events is stressed; but 4.1 (with or without Hecate) is less urgent in presentation and less specific in time (despite Macbeth's intention in 3.4.133-4 to consult the Sisters 'betimes'). Act 4, Scene 3, marks a complete change of pace; it is often referred to as long, and indeed seems so in the theatre; but it is not actually as long as it feels, it is the pace which has been changed. After that, an audience would be hard put to say how soon the events of Act 5 followed those of Act 2. Theatrical time depends very much on stage rhythm, and Shakespeare often used Act 4 of a play to alter the perspective on time in the whole, thus gaining an overall amplitude for an action which has previously been compressed. *Hamlet* loses its urgency when the Prince goes meekly off to England, and when editors used to impose a realist's calendar on the plays, *Othello* was said to have a 'double time-scheme'.

Shakespeare derived the dialogue between Malcolm and Macduff from Holinshed's second major digression. It is a rhetorical exercise of a type Holinshed frequently borrowed from other sources, but seldom originated for himself; where he found it I do not know; it is a skilful and sophisticated debate in itself, but its very difference of form clogs the narrative, and it is precisely that quality which Shakespeare utilizes to effect a change of pace and therefore of perspective. The battle of Act 5 is greatly compressed from a long and opportunist campaign in Holinshed which is only turned to Malcolm's advantage finally by the arrival of the English force. Boethius was by no means pleased by that, and blamed the Anglicized Malcolm for the moral deterioration of the Scots, in a passage which Holinshed transferred to his initial 'Description' of Scotland: by contact with the English, they began 'to learn also their manners, and therewithal their language ... the temperance and vertue of our ancestors grew to be judged worthy of small estimation amongst us, notwithstanding that a certain idle desire of our former renown did still remain with us' (p. 507). This judgement colours the account of Malcolm's reign, and contributes presumably to the equivocal figure that Shakespeare makes of him; it also contributed Macbeth's ironic 'Then fly false thanes, | And mingle with the English epicures' (5.3.7–8).

(ii) STUART POLITICS

The Macbeth story was used to provide the basis for the Stuart myth, which is an obvious reason for choosing it. In 1606 William Warner produced a *Continuance* of his *Albion's England* (first published in 1586) in which he included Macbeth and traced the Stuart claims to both thrones. The interest in a new king and a new dynasty is obvious. But that was there while Elizabeth still lived (see p. 59), and it is a dangerous step to assume that because the play responded to public interest it was deliberately designed to flatter the King. James was a more active patron than Elizabeth had ever been, and took over the Lord Chamberlain's Men as the King's Men. That obliged them to give periodic performances at court, and also to provide professional assistance for the largely amateur productions of court masques. In other respects there is no evidence of positive intervention by the King, but plenty that is negative, of the Master of the Revels exercising a sensitive political censorship. In 1604 a play about the Gowrie conspiracy to kill James in Edinburgh, which would have served a similar interest, was suppressed altogether. Its particular offence is not known, but the representation of a reigning monarch was never allowed. Discretion was clearly necessary, but active flattery is quite another matter.

In the late eighteenth century, Malone suggested that *Macbeth* might have been one of the three plays performed before the court when James's father-in-law King Christian of Denmark paid a state visit to London in 1606;[1] but the plays are not named in the accounts, and Malone did not pretend that there was any evidence to support his conjecture. None the less, the idea has frequently been revived, especially since the publication of H. N. Paul's *The Royal Play of Macbeth* in 1950, and it is sometimes treated now as if it were a known fact. The belief needs rebuttal, for it not only becomes a false evidence for dating the play, but also distorts critical judgement of it. Paul brought together a considerable amount of interesting historical material and was scrupulous in handling its detail; but for his central thesis, that the play was written by James's command for performance at Hampton Court in August 1606, he had no evidence whatsoever. He invented, first, an imaginary meeting between Shakespeare and James in Oxford during a royal visit in 1605, where Shakespeare was supposed to have been on account of his (legendary) friendship with an innkeeper and his wife (the parents of William Davenant the dramatist). No serious argument is put forward to support this, but thereafter Paul refers to it as if it had been proven. The mass of material that follows, however useful in itself, does nothing to support the fantasy. Paul himself pointed out that what is superficially the strongest internal support for his idea, the description of Edward the Confessor 'touching' for the King's Evil (scrofula) in 4.3.140–59, would not have been safe flattery. James did indeed touch occasionally, but with great reluctance, and serious theological doubts caused him to modify the ceremony; he expressly rejected the suggestion of miracle, which Shakespeare retained in l. 147. To cope with this, Paul added another guess, that the Council was eager for James to continue touching despite his scruples, and so used the Master of the Revels (Sir George Buck) to see that Shakespeare interpolated the passage, not as flattery but as persuasion of the King. There is no evidence for this, any more than for the claim (p. 41) that 'Everything points to the performance at Hampton Court on August 7 as that of *Macbeth*'; instead, we are told that 'the internal forces can only be known by peering into the laboratory

---

[1] E. K. Chambers, *The Elizabethan Stage* (Oxford, 1923), iv. 173.

of the dramatist's mind as he did his work' (p. 43); a trick that is extended into the minds of James, Buck, and the rest. A historian who offers a profound study of a man may properly use his insight to interpret known actions; but Paul offers no such study, and offers his insight only to invent actions which are not known at all.

It is, indeed, by no means clear that the play sustains flattery of the King. Attention is not focused on the political theory of kingship in the way that it had been in earlier plays, from *Richard II* to *Julius Caesar*. All the significant figures, who might have pointed to James, are sooner or later involved in equivocal judgement; the only exception is Edward the Confessor, who does not appear on stage, and is in any case an English not a Scottish king. Compared with Holinshed, Shakespeare may seem to have idealized Duncan, but age is the only particular virtue with which he has been endowed and not all the strictures are forgotten; one of the play's conspicuous ironies is his comment on the disgraced Cawdor:

> There's no art
> To find the mind's construction in the face;
> He was a gentleman on whom I built
> An absolute trust.   (1.4.11–14)

He is immediately brought face to face with the new Cawdor, so hastily promoted, whom the audience has just heard meditating regicide (1.3.135–43). Banquo, supposedly James's ancestor, is allowed a general honesty, but it is by no means so clear as Horatio's in *Hamlet*.

The show of eight kings in 4.1 explicitly dramatizes Banquo's royal descendants, but in a slightly odd fashion: James himself was the ninth Stuart monarch, and his immediate predecessor was his mother Mary, who was queen in her own right (his father, Lord Darnley, himself a Stuart, was proclaimed king, but only in his wife's right). After the murder of Banquo, Holinshed enumerates the full genealogy of the Stuarts (172/b/61–174/a/25; not given in detail in Bullough, p. 499); he does not here enumerate the monarchs, and though they could be counted, their number is obscured by the listing of numerous siblings and their spouses. Sixty pages later Holinshed remarks 'Thus ye may perceive how the Stewards came to the crown, whose succession have enjoyed the same to our time: queen Mary mother to Charles James that now reigneth being the eighth person from this Robert' (the Second, first Stuart king) (245/a/67-74, not in Bullough). The sentence could easily be misconstrued as meaning James was the eighth, though carefully read it clearly means Mary, who was still alive when Holinshed wrote. Much later in the *Chronicles* (390/b/1-49, not in Bullough), but in a passage clearly indicated in Holinshed's own index under a heading about under-age succession, he notes that after the first two, all Stuart kings had been under age at their accession; this time he lists them and numbers them from one to nine, and the ninth is unambiguously James. Holinshed was impressed by this because under-age succession is an obvious practical disadvantage to the principle of primogeniture and one that he repeatedly points out as a reason why it was not a decisive claim in early Scottish history, despite several efforts to establish dynasties. Malcolm, in *Macbeth*, is made technically of age when Duncan creates him Prince of Cumberland (1.4.40); but Macbeth's consternation makes it clear that Malcolm is young for the promotion. Malcolm himself states in 4.3.14 'I am young', and in 5.3.3 Macbeth refers to him contemptuously as 'the boy Malcolm'. Malcolm is initially dispossessed because he flees, but this occasions no great surprise or regret among the thanes, and in 2.4.29-32 it is suggested that the Scottish monarchy was actively elective.[1] Holinshed's interest in this topic might well have suggested the symbolic apparitions in 4.1: an armed head (the initial usurpation), a bloody child, a child crowned with a tree in his hand, and finally the show of eight kings.

'King' was occasionally used of a queen regnant, but the sense is rare, and they are traditionally presented on the stage as all male (see Fig. 7). Mary was a controversial figure, but James would not have been flattered by her omission. If 'eight' was not a mere slip, it might be that omitting the ninth was to avoid even symbolically representing the reigning monarch; but if all were male on stage before the Restoration there would still be inconsistency. Whatever the explanation, it seems that the public myth is being stressed rather than personal flattery of the King.

That is the conclusion of an important essay by a historian,

---

[1] Later, the elective element in the succession to European monarchies (except the Holy Roman Empire) became little more than a formality.

Michael Hawkins, on 'History, Politics, and *Macbeth*'.[1] Hawkins provides a useful summary of the central political concerns of the sixteenth and early seventeenth centuries (pp. 160–3) and finds their representation in the play to be always ambiguous. They are: family loyalty, historically typical of pre-feudal societies; extended to the more complex bonds of feudalism; and later narrowing towards a stress on central government in the person of the monarch (developing a bureaucracy which does not arise in the play). Family loyalty in its pre-feudal form is appropriate to eleventh-century Scotland; contrary to general belief, it is characteristic of Shakespeare's history plays, English and Roman, to show an interest in other societies and governmental structures, as well as using them to reflect, and sometimes to question, the assumptions of his own. Julius Caesar's attempt to convert the Roman republic into a monarchy is seen primarily as dangerous for Rome, but it also illuminates the virtues of republican institutions; the balance between symbolic kingship and practical competence in *Richard II* was dangerously uncertain so that Act 4 was censored; election and designation triumphs over heredity in *Hamlet*; and so on.

Family bonds recur in *Macbeth*, but never finally dominate. The Macbeths are childless; Malcolm is young and uncomfortably callow; Macduff is motivated as much by bloody revenge as by concern for the state of Scotland. Shakespeare's respect for the 'otherness' of place and time means that feudal bonds do not become explicit, but he follows Holinshed in giving the thanes the chivalric virtues of courage, loyalty, honour, and goes beyond Holinshed in providing them with a language of courtesy whose potential for equivocation I have discussed (pp. 21–2). Both these political modes are presented, as Hawkins insists, ambiguously; the third level, personal monarchy, is even more so. James, in *Basilikon Doron*, recognized the inevitable conflict between private and public morality, as well as the necessity for political expediency.[2] Duncan's age makes him a patriarchal king, but his virtues are essentially private, not public, and of expediency he has no trace. Even more telling is the problem of tyranny: Hawkins remarks (pp. 176–7) that James argued 'one strand of orthodox opinion ... that the worst tyrant's oppression should be willingly accepted since it was sent by God to test the Christian

[1] In John Russell Brown, ed., *Focus on Macbeth* (1982), 155–88.
[2] Ibid. 173.

humility of the subject'. Macbeth is primarily attacked in the play as a tyrant, because the thanes are never publicly certain of his guilt in murdering Duncan—and his assumption of the throne was not otherwise a usurpation. This issue, too, remains ambiguous and is not resolved in favour of James's well-known views. Hawkins concludes (p. 180) that all in the play 'operate in partial darkness in a very volatile political world'; the politics of the play therefore are involved in its ambiguity and darkness in other respects, and its response to James is far from simple flattery. Hawkins is surely right to insist that it would have been an unlikely choice for presentation to the two kings.

The myth of the Weïrd Sisters prophesying to Banquo the lineage of the Stuart dynasty was presented in a short, appropriately fulsome, Latin dialogue when James visited Oxford on 19 August 1605. It is given, with a translation, in Bullough (p. 470), and it is unambiguous flattery. It adds nothing pertinent to the play that was not in all the chronicles; Shakespeare may have heard of it, but it was not printed until 1607, and he certainly did not echo it.

My conclusion is, therefore, that the choice of subject had far more to do with public interest, and therefore the public theatre, than with pleasing the King. Indeed, the subject had obvious political dangers, and it is more probable that the material cited by Paul was motivated by the need to avoid censorship than by flattery. Stuart politics are certainly an important source for *Macbeth*, but not a royal command.

(iii) CLASSICAL LADIES

It has often been suggested that phrases from Seneca suggested images and even rhetorical structures in *Macbeth*; more so, it seems, than in most of Shakespeare's plays except for *Titus Andronicus* and *Richard III*. The connection may not be coincidence: Seneca's heroes are either evil in themselves or, like Hercules, become so in a fit of madness; Titus becomes insanely committed to evil, Richard was always so, and Macbeth descends into the physical evil he first perceived in his own mind. Shakespeare probably read Seneca in Latin at school, and seems also to have known the Tudor translations; stray phrases might as easily have come from memory as from a deliberate re-reading such as Muir (*Sources*, p. 214) suggested.

In fact, it seems to be for Lady Macbeth that he used Seneca, rather than for her husband. The plays most often suggested are *Medea* and *Agamemnon*. In both, powerful women, Medea and Clytemnestra, take revenge on their husbands, and in both children are involved: Agamemnon had sacrificed his and Clytemnestra's daughter Iphigenia to secure a fair wind to Troy; Medea killed the children she had by Jason when he deserted her. Both might have contributed to the development of Lady Macbeth from Holinshed's account of Donwald's wife, and to the phantom children she offers as part of her femininity; but, as Inga-Stina Ewbank pointed out, Medea is by far the closer.[1] Medea invokes Hecate, mixes a hellish brew (suggesting the Weïrd Sisters more than Lady Macbeth), and asks to be 'unsexed' before she murders her children:

> If ought of ancient courage still do dwell within my breast,
> Exile all foolish female fear, and pity from thy mind . . .

And shortly afterwards

> since my womb hath yielded fruit, it doth me well behove
> The strength and parlous puissance of weightier ills to prove

> How wilt thou from thy spouse depart? As him thou followed hast
> In blood to bathe thy bloody hands . . .

With Medea this describes facts; with Lady Macbeth it relates to a concept of femininity and an imaginative rejection of it.

Clytemnestra's strength of personality makes Agamemnon relatively unimpressive; Medea's husband, Jason, is explicitly feeble in his betrayal of her to a more conventional life. Both women become violently destructive, but their tragedies depend on their husbands' failures; Lady Macbeth's need to bolster up a vacillating man has striking similarities, and it is of course Macbeth who first displays the ambition which she sustains. If she has less provocation than her classical sisters, it should be said that she is also less successfully destructive: they do kill, as she is unable to (2.2.13–14). Nor is she offered as a type of universal female evil, as is sometimes suggested and as Holinshed might

---

[1] 'The Fiend-like Queen: A Note on *Macbeth* and Seneca's *Medea*', *Shakespeare Survey* 19 (1966), 82–94, reprinted in Kenneth Muir and Philip Edwards, eds., *Aspects of Macbeth* (Cambridge, 1977), 53–65. Quotations are from Studley's translation of *Medea* in *Tenne Tragedies* (1581), 2 vols., Tudor Translations (1927), ii. 53–98.

easily have seemed to imply: her offence depends on 'unsex me here' (1.5.40), a reversal primarily of her humanity, secondly of her femaleness. Equally, it is not true that her opposite in the play is woman as perfect image: the brief cameo of Lady Macduff is attractively spirited and, goaded by her child, she exhibits a refreshing independence; she is certainly not an icon on a pedestal. Finally, when Lady Macbeth herself asserts the male–female antithesis by continually challenging Macbeth's 'manhood', he retorts tellingly 'I dare do all that may become a man, | Who dares do more is none' (1.7.46–7). As much as the violence of Medea is moderated into partial fantasy with Lady Macbeth ('I have given suck', 1.7.54), so also the male–female antithesis is moderated; her 'unsex me here' is partly ambiguous, and Macbeth's lines play on 'man' as male and on man as mankind.

(iv) WEÏRD SISTERS

Lady Macbeth's sorcery seems to be purely verbal; the Weïrd Sisters are another matter. The persecution of 'witches' studied in Keith Thomas's *Religion and the Decline of Magic* makes it abundantly clear that crude misogyny played a large part in their trials. But the subject was controversial at the time: the best-known Elizabethan study was Scot's *Discovery of Witchcraft* (1584), which consistently attacked belief in witchcraft and stressed the personal revenges on which the evidence depended. King James, then still in Scotland, retorted in print with *Demonology* (1597); he ardently attacked Scot, and related traditional beliefs to his own experience. A pamphlet printed in 1591, *News from Scotland*, records the trial of Agnes Sampson and others, and her 'confession' that she had attempted to poison the King as well as listening in to his private thoughts.[1] When he reached London in 1603, James ordered the reprinting of *Demonology* there, and induced parliament to re-enact and strengthen the law against witchcraft. He had personally attended Agnes Sampson's trial in Berwick, and continued the practice of going incognito to trials in

---

[1] 'And thereupon taking his Majesty a little aside, she declared unto him the very words which passed between the king's Majesty and his Queen at Upslo in Norway [?Uppsala, in Sweden] the first night of marriage, with their answer to each other', *Newes from Scotland* (1591), facsimile ed. H. Freeling for the Roxburgh Club, 1816, B2^r–B2^v. The short pamphlet is very vague about the trial but offers detailed, often pornographic, descriptions of witchcraft and torture. The historical charge involved the Earl of Bothwell.

England; but it seems that his interest shifted towards exposing false charges and therefore quashing convictions; still, if his confidence weakened, the Act of 1603 was not repealed.

There is no evidence at all that Shakespeare made use of any of these works, either in *Macbeth* or in his early plays on Henry VI's reign where both Joan of Arc and the Duchess of Gloucester are shown summoning powers of darkness. I have cited *Demonology* in commentary on 1.3.124–7, and *News from Scotland* for 4.1.6–8, but in both cases only to illustrate common beliefs, not to suggest verbal borrowings. Scot certainly seems to be involved with the song in 4.1, but that is not supposed to be Shakespeare's (see pp. 57–9, 66).

I have already discussed (pp. 2–4) the relation between village witchcraft and goddesses of destiny in the Weïrd Sisters. All that Shakespeare used of the first was common knowledge, including the familiars and the creatures that go into the cauldron in 4.1; they go back to classical as well as medieval sources, to Ovid as well as Seneca's *Medea*; but though the brews are similar, they are not close enough to suggest copying. For the Weïrd, no source has been suggested, beyond Holinshed's hints; no doubt this was eked out with recollections of classical mythology, and especially the oracular ambiguities of the Delphic and other sybils. The result is an amalgam which seems to be wholly original, as Anthony Harris remarked in *Night's Black Agents* (Manchester, 1980): 'as in *A Midsummer Night's Dream* Shakespeare revolutionised the centuries-old view of fairies ... so in the Weïrd Sisters he produced a unique creation' (pp. 43–4).

(v) DEVIL PORTERS

The Porter's solo performance in the first nineteen lines of 2.3 is based on the miracle plays of the fourteenth century. The traditional cycles disappeared gradually during the sixteenth century, but survived in popular memory long after the reformed church finally put an end to their performance. None of the surviving texts has a designated porter on the mouth of Hell, but various plays dealing with Christ's activities after his death offer two or three devils on the gate, alarmed at first by Christ's arrival but later happily ushering in the damned. The 'Descent into Hell' in Chester and in the *Ludus Coventriae* leads on to the 'Harrowing of Hell' in the *Ludus Coventriae* and in York, and so to the

'Last Judgement' plays in Chester, York, and the Towneley cycle (here a version of the York play). Macbeth's Porter seems to combine that role with a parody of St Peter's function at the gate of Heaven, since in the plays the damned do not seek their admission into Hell. In the York 'Harrowing of Hell' (ed. L. Toulmin Smith, Oxford, 1885) it is Jesus who demands admission:

> *Principes, portas tollite,*
> Undo your gates, ye princes of pride
> *Et introibit rex glorie,*
> The king of bliss comes in this tide.   (ll. 181–4)

In the Last Judgement plays, especially at Chester, the devils welcome many a social oppressor as the Porter does in *Macbeth*.

A minor puzzle has been found in his behaviour with names: he invokes the name of Beelzebub in line 3, but in line 7 affects anonymity with 'Who's there in th' other devil's name?' Beelzebub is one of the very few devils named in the Bible; in Matthew 12: 24–7 he is named as 'the prince of devils' and Jesus apparently identifies him with Satan; for the Porter he is one of the diabolic trinity headed by Lucifer, as in the Last Judgement plays imitated by Marlowe in *Dr. Faustus*, where Lucifer enters the final act attended by Beelzebub and Mephistophilis. The Porter's ignorance of 'th' other devil's name' has usually been attributed to a failure of memory, but it is more likely to derive from the anonymity usual for the attendant devils in the miracle plays. At York, Lucifer acts alone in Heaven, but after his fall he emerges from Hell-mouth with a companion who is not named; in Chester they are together in Heaven as Lucifer and Lightburn, which is simply an Anglicization of Lucifer (Marlowe used that too, in the last act of *Edward II*, where Lightborn is the mysterious joker who murders the King—seemingly a reincarnation of Edward's male lover, Gaveston: buggery was, and all too often still is, associated with the devil). In Hell, the Chester devils become anonymous, though the dialogue makes it clear they are the same pair that were in Heaven; Towneley is similar, but less clearly defined. The minor devils in the Last Judgement plays are all anonymous, though the princes are sometimes named: Satan in Chester; Beelzebub, Satan, and Belial in York; Titivillus in Towneley; Belial in the *Ludus Coventriae*. In the Newcastle Noah play, Diabolus refuses to answer Mrs Noah's request for his name. The insistent anonymity clearly relates to a tradition, probably connected with the popular fear of naming the devil. This is surely the point of the Porter's joke.

EDITORIAL PROCEDURES

THE text of the play is based, as accurately as possible, on the Folio of 1623, modernized according to the principles expounded in Stanley Wells, *Modernizing Shakespeare's Spelling* (Oxford, 1979), and in Gary Taylor's edition of *Henry V* (Oxford, 1982), 75–81. Old spellings, however interesting, are not reliable indicators of pronunciation and their retention makes an impact more quaint than enlightening to the modern reader; in a few cases the change of form has been more substantial, and these are represented in the collation thus:

account] F (accompt)

Punctuation presents a different problem: modern practice must be respected to the degree that reading the text will not be disturbed by mere oddity; but modern punctuation is still rigorously grammatical in theory, and sometimes has to be simply disregarded in speech; it may help to disentangle complex structures, and yet hinder the reader or, especially, the actor looking for the rhythm and stress of the language. Seventeenth-century punctuation paid more regard to rhetorical values. Folio punctuation is certainly not Shakespeare's—compositors then, as now, had their own methods and applied them at will; but it does sometimes offer useful indicators which can be retained without, I hope, confusing the reader. All punctuation is, and must be, a compromise; all that I have done is to incline slightly more to seventeenth-century practice than other editors prefer; in particular, I have tried not to introduce grammatical stops that interfere with the rhythm of speech. Changes, however, from the Folio punctuation are necessarily frequent, and are only recorded where F is ambiguous and the modern form is not. The final -ed of past participles is always printed as in modern English; it is to be sounded or silent in the usual way except where the rhythm demands a syllable nowadays silent, and then it is marked with an accent, èd; occasionally it may or may not be sounded, and then it is left to individual choice.

Speech prefixes are given in full and are silently normalized where F shows variety. The stage directions in *Macbeth* are

83

generally believed to derive from Shakespeare via erratic modifications by the book-keeper in the theatre. They are generally retained here as an important, though sadly inadequate, record of practice in the theatre, but modified and augmented where necessary; doubtful points of staging are discussed in the commentary. All changes are recorded in the collation, but those where the action appears to be certain are not distinguished in the text; where there is any doubt, added or altered words are enclosed in partial square brackets: ⌈ ⌉. All indications for asides are editorial; the Folio never indicates who is speaking to whom.

Act and scene divisions are indicated for reference in the margin, based on those in the Folio where they are in Latin (*Actus Primus. Scoena Prima . . . Scena Secunda . . . Scena Tertia,* etc.); these are not noted in the collation. The act divisions would not have been intended to show in performance, which was continuous at the Globe; they may, however, represent accurately a structural plan, however invisible, just as renaissance architectural drawings basing the design of cruciform churches on the proportions of the human body could never have been intended to be perceived by users of the buildings that resulted. Scene divisions were based on the 'English' practice, marked whenever there is a clear stage, but F shows the battle in Act 5 as continuous in 5.7, where most editors have made further divisions—without agreeing on the number needed: Muir has altogether nine scenes in Act 5, Oxford eleven. I have followed F, because the action is a continuous representation of the battle although only individual combatants are actually seen. The locations invented by editors with scenic theatres, or novels, in mind have been entirely ignored; location is indicated in dialogue on the rare occasions when it is relevant (e.g. 4.3 in the English Court), and there was never any justification for such theatrical impossibilities as 'A plain before the castle', 'Another part of the plain', 'Another part of the field' (5.6 Rowe, 5.7 Capell, 5.8 Dyce, as given in Muir) for successive very brief 'scenes' in Act 5.

There is no cast list in the Folio for *Macbeth*, as there is for some plays; the rather formidable list of 'Persons of the Play' is given here in accord with the usual practice of setting out all roles named or specified ('servants', 'soldiers', etc.) in stage directions or text, but this should not be understood as indicating the size of the company required to present them, still less that the King's Men were anything like so numerous. Doubling was standard practice, and significant roles appearing in only a single scene may well call for experienced actors; there are several such in Macbeth (e.g. Captain, Porter, Old Man, etc.). The limits to possible doubling are set by the fact that persons who are ever on stage together cannot be doubled, and further by the fact that in continuous performance no actor can leave the stage at the end of one scene and immediately re-enter the next, whether in the same role or a new one. I calculate that Macbeth could be satisfactorily performed by a company of twelve, of whom no more than eight need to be experienced—and indeed successful touring companies have demonstrated this in recent years. For the last act most of the earlier roles have disappeared and several new ones are introduced; as usual in Shakespeare battles are largely off-stage, 'armies' represented largely by brief collocations of leaders. It is likely enough that the theatre employed a number of stage-hands who were not actors but could be used to swell a crowd of extras, whether of servants or of soldiers, but both could be supplied by actors in simple robes in this play.

In some plays there is reason to think that doubling had special significance, requiring that the actor be recognized, though usually convention would overlook his identity—just as male actors playing women were usually accepted, but the convention could be called attention to (in As You Like It, Twelfth Night, or Jonson's Epicoene, for instance). Marlowe seems to have experimented with significant doubling in both Tamburlaine and Edward II. In Shakespeare, the symmetry of A Midsummer Night's Dream clearly suggests doubling Theseus with Oberon and Hippolyta with Titania, and the ending is arranged to make that, magically, possible. In Lear, the Fool appears only after Cordelia's exit from Act 1, and disappears without explanation before her re-entry in Act 4; between them they span the range of 'Folly' analysed by Erasmus, and their doubling is hinted in Lear's ambiguous 'And my poor fool is hanged' (Lear 5.3.281). The scope for suggestive doubling seems to be considerable in Macbeth, especially with the Weïrd Sisters, but I see very little specific evidence for it except for one new figure in Act 5. If, as I suppose, Seyton was pronounced like 'Satan', his sudden appearance in 5.3 (after three calls from Macbeth) as an amanuensis Macbeth has never had before suggests association with the First Witch or even, possibly, with Lady Macbeth; in 5.5 he displays a knowledge of the Queen's death without, in the Folio, leaving the stage—a strange irony if they were played by the same actor (Lady Macbeth was last seen in 5.1). That is only a tentative suggestion; it is more probable that the slightly odd inclusion and exclusion of Lennox and Ross in stage directions for Act 5 relates to the use of reliable actors for other roles hereabouts: Lennox is named only for 5.2, and Ross does not appear until the very end of 5.7. Hence Lennox could easily play Seyward, for instance, and Ross the Doctor and/or Young Seyward.

The collation below the text shows all verbal departures from F; the edition cited immediately after the lemma is that in which the reading first appeared; a bracketed entry records that an earlier scholar had suggested the reading but not himself adopted it; where an emendation which has found, or might plausibly find, acceptance has not been adopted here, it is also collated. Lineation creates special problems in *Macbeth* and would overload the main collation: doubtful lines are fully recorded and discussed in Appendix A, and important cases noted in the commentary on the page.

Line-numbers and quotations from other Shakespeare plays are taken from *The Complete Works*, edited by Stanley Wells and Gary Taylor (Oxford, 1986). References to other works printed before 1660 are to old-spelling editions; quotations from them have been modernized here in both spelling and punctuation.

The commentary tries to clarify meanings without reducing possible ambiguity, often by listing separate senses for words which may all (in varying degrees) be relevant in the context; the primary sense, if there is one, given first. It also discusses textual and performance problems where they arise. Shakespeare often used, or alluded to, proverbs and similar sententious phrases; since this registers a significant element in his language which may not always be recognized, I have frequently called attention to it even where literal understanding is not affected, but the number of possible cases is so large that I have omitted several that seem insignificant. Some phrases only became proverbial after he had coined them and these are not noted. M. P. Tilley's *Dictionary* (1951) is the primary source of information; all his entries for Shakespeare are reprinted in Dent's *Index* (1981), using Tilley's reference system, with some additions (marked by decimal points), and a few deliberate deletions. I have therefore given references for economy to Dent alone.

Abbreviations and References

Place of publication is London unless otherwise stated. Titles and quotations from early texts are normally given in modern spelling.

EDITIONS OF SHAKESPEARE

Seventeenth-century texts are in chronological order; later editions are alphabetical.

F, F1	The First Folio, 1623 (F1 is used when necessary to distinguish it from later Folios)
Drexel	Drexel MS 4175, New York Public Library, Lute Songs, *c*.1630
Bull	Bull Manuscript, MU. MS 782, Fitzwilliam Museum, Cambridge (England), Lute Songs, *c*.1630
Witch	Thomas Middleton, *The Witch*, Malone MS 12, Bodleian Library, Oxford, *c*.1630; ed. W. W. Greg and F. P. Wilson, Malone Society Reprints (Oxford, 1948/50)
F2	The Second Folio, 1632
F3	The Third Folio, 1663
1673	*Macbeth*, a Tragedy acted at the Dukes Theatre, 1673
Yale	*Davenant's Macbeth from the Yale Manuscript*, ed. Christopher Spencer (New Haven, 1961)
1674	*Macbeth*, a Tragedy: with all the alterations, amendments, additions, and new songs. As it is now acted at the Dukes Theatre, 1674
F4	The Fourth Folio, 1685
Boswell	James Boswell, ed., *The Plays and Poems of William Shakespeare*, 21 vols. (1821)
Cambridge	W. G. Clark and W. A. Wright, eds., *The Works of William Shakespeare*, The Cambridge Shakespeare, 9 vols. (Cambridge 1863–6)
Capell	Edward Capell, ed., *Mr William Shakespeare His Comedies, Histories, and Tragedies*, 10 vols. (1767–8)

Clarendon	W. A. Wright, ed., *Macbeth*, Clarendon Press Series (Oxford, 1869)
Collier	John Payne Collier, ed., *The Works of Shakespeare*, 8 vols. (1842–4)
Delius	N. Delius, ed., *Shakespeares Werke*, 2 vols. (Elberfield, 1869)
Foakes	R. A. Foakes, ed., *The Tragedy of Macbeth*, The Bobbs-Merrill Shakespeare Series (Indianapolis/New York, 1968)
Globe	W. G. Clark and W. A Wright, eds., *The Works of William Shakespeare*, The Globe Edition (1864)
Hanmer	Thomas Hanmer, ed., *The Works of Mr William Shakespear*, 6 vols. (Oxford, 1743–4)
Hudson	H. N. Hudson, ed., *The Works of Shakespeare*, 11 vols. (Boston, 1851–6)
Hunter	G. K. Hunter, ed., *Macbeth*, The New Penguin Shakespeare (1967)
Johnson	Samuel Johnson, ed., *The Plays of William Shakespeare*, 8 vols. (1765)
Keightley	Thomas Keightley, ed., *The Plays of William Shakespeare*, 6 vols. (1864)
Knight	Charles Knight, ed., *The Pictorial Edition of the Works of Shakespere*, 8 vols. (1838–43)
Malone	Edmond Malone, ed., *The Plays and Poems of William Shakespeare*, 10 vols. (1790)
Moberly	C. E. Moberly, ed., *Macbeth*, Rugby edition (1872)
Muir	Kenneth Muir, ed., *Macbeth*, The Arden Shakespeare (1951, rev. 1984)
Oxford	Stanley Wells and Gary Taylor, eds., *The Complete Oxford Shakespeare* (Oxford, 1986). In modern spelling; see also the companion volume in original spelling (1986), and *Textual Companion* (1987)
Pope	Alexander Pope, ed., *The Works of Mr. William Shakespear*, 6 vols. (1723–5)
Pope 1728	Alexander Pope, ed., *The Works of Mr. William Shakespear*, 10 vols. (1728)
Riverside	G. Blakemore Evans, ed., *The Riverside Shakespeare* (Boston, 1974)
Rowe	Nicholas Rowe, ed., *The Works of Mr William Shakespear*, 6 vols. (1709)

Rowe 1714	N. Rowe, ed., *The Works of Mr William Shakespear*, 3rd edn., 8 vols. (1714)
Singer	S. W. Singer, ed., *The Dramatic Works of William Shakespeare*, 10 vols. (Chiswick, 1826)
Singer 1856	S. W. Singer, ed., *The Dramatic Works of William Shakespeare*, 10 vols. (1856)
Steevens	S. Johnson and G. Steevens, eds., *The Plays of William Shakespeare*, 10 vols. (1773)
Steevens 1778	Samuel Johnson and George Steevens, eds., *The Plays of William Shakespeare*, 10 vols. (1778)
Steevens 1793	George Steevens and Isaac Reed, eds., *The Plays of William Shakespeare*, 15 vols. (1793)
Variorum	H. H. Furness, ed., *Macbeth*, A New Variorum Edition of Shakespeare, vol. ii (Philadephia and London, 1901–3)
Warburton	William Warburton, ed., *The Works of Shakespear*, 8 vols. (1747)
White	R. G. White, ed., *The Works of W. Shakespeare*, 12 vols. (Boston, 1857–66)
Wilson	J. Dover Wilson, ed., *Macbeth*, The New Shakespeare (Cambridge, 1947)

OTHER WORKS

Abbott	E. A. Abbott, *A Shakespearian Grammar*, 2nd edn. (1870)
Boethius	Hector Boethius, *Scotorum Historiae* (Paris, 1527); Scottish translation by John Bellenden, *The Chronicle of Scotland* (?1536), ed. E. B. Batho and H. W. Husbands, Scottish Text Society (Edinburgh, 1941)
Bullough	Geoffrey Bullough, ed., *Narrative and Dramatic Sources of Shakespeare*, vii. *Major Tragedies* (London and New York, 1973)
Demonology	King James I, *Demonology* (Edinburgh, 1597; London, 1603)
Davenant	William Davenant, *Macbeth*: see Yale and 1674
Dent	R. W. Dent, *Shakespeare's Proverbial Language: An Index* (Berkeley, 1981)
Downes	John Downes, *Roscius Anglicanus, or an Historical Review of the Stage: After ... 1660* (1708); ed. J. Loftis (Los Angeles, 1969)

Empson	William Empson, *Seven Types of Ambiguity* (1930; 3rd edn., 1953)
Hakluyt	Richard Hakluyt, *The Principle Navigations Voyages Traffiques and Discoveries of the English Nation*, 2nd edn., 1598–1600, 12 vols. (Glasgow, 1903–5)
Hawkins	Michael Hawkins, 'History, Politics, and *Macbeth*', in J. R. Brown, ed., *Focus on Macbeth* (1982), 155–88
Hibbard	Howard Hibbard, *Bernini* (1965)
Hinman	Charlton Hinman, *The Printing and Proof-Reading of the First Folio of Shakespeare*, 2 vols. (Oxford, 1963)
Holinshed	Raphael Holinshed, *The Chronicles of England, Scotland, and Ireland*, 2 vols., 2nd edn. (1587), vol. i
Inchbald	Mrs Elizabeth Inchbald, 'Remarks', in *Macbeth* ... As performed at the Theatres Royal, Covent Garden and Drury Lane. Printed under the authority of the Managers from the Prompt Book (1808), pp. 3–4
Lake	D. J. Lake, *The Canon of Thomas Middleton's Plays* (1975)
News from Scotland	*Newes from Scotland* (1591); facsimile ed. H. Freeling, The Roxburgh Club (1816)
OED	*Oxford English Dictionary* (1st edn.)
Paul	H. N. Paul, *The Royal Play of Macbeth* (New York, 1950)
Pepys	Robert Latham and William Matthews, eds., *The Diary of Samuel Pepys*, vols. v–viii (1971–4)
Purchas	Samuel Purchas, *Hakluytus Posthumus or Purchas his Pilgrimes*, 1625, 20 vols. (Glasgow, 1903)
Record	W. W., *A true and just Recorde, of the Information, Examination and Confessions of all the Witches, taken at S. Oses in the Countie of Essex* (1582)
Scot	Reginald Scot, *The Discovery of Witchcraft* (1584); ed. Brinsley Nicholson (1886)
Shirley	Frances A. Shirley, *Shakespeare's Use of Off-stage Sounds* (Lincoln, Nebr., 1963)
Thomas	Keith Thomas, *Religion and the Decline of Magic* (1971)
Tilley	M. P. Tilley, *A Dictionary of the Proverbs in England in the Sixteenth and Seventeenth Centuries* (Ann Arbor, Mich., 1950)

Macbeth

THE PERSONS OF THE PLAY

FIRST WITCH ⎫
SECOND WITCH ⎬ the Weïrd Sisters
THIRD WITCH ⎭

HECATE, Goddess of the Witches
THREE OTHER WITCHES, singers and dancers with Hecate
APPARITIONS

MACBETH, Thane of Glamis, then of Cawdor, later King
LADY MACBETH, later Queen
DUNCAN, King of Scotland

MALCOLM ⎫
⎬ Duncan's sons
DONALBAIN ⎭

BANQUO ⎫
FLEANCE, Banquo's son
MACDUFF, Thane of Fife
LADY MACDUFF
SON of the Macduffs
LENNOX ⎬ thanes (Scottish lords)
ROSS
ANGUS
MENTEITH
CAITHNESS ⎭

CAPTAIN or Sergeant
PORTER of Macbeth's Castle
OLD MAN
THREE MURDERERS (of Banquo)
LORD (anonymous, in 3.6)
DOCTOR (English)
DOCTOR (Scottish)
GENTLEWOMAN (Nurse)
SEYTON, Macbeth's armourer

SEYWARD, Earl of Northumberland ⎫
⎬ English lords
SEYWARD'S SON ⎭

Various Attendants, Messengers, Servants, a Sewer, Murderers (of Lady Macduff), Soldiers of two armies—Malcolm's and Macbeth's

Macbeth

1.1   *Thunder and lightning.*
       Enter three Witches

FIRST WITCH

When shall we three meet again?
In thunder, lightning, or in rain?

SECOND WITCH

When the hurly-burly's done,
When the battle's lost, and won.

THIRD WITCH

That will be ere the set of sun.

FIRST WITCH

Where the place?

SECOND WITCH   Upon the heath.

THIRD WITCH

There to meet with Macbeth.

FIRST WITCH I come, Graymalkin.

SECOND WITCH Paddock calls.

THIRD WITCH Anon.   10

ALL

Fair is foul, and foul is fair,
Hover through the fog and filthy air.   *Exeunt*

1.1.9–10 SECOND WITCH ... Anon] SINGER 1856; *All. Paddock* calls anon F

1.1.0.1 **Enter three Witches** Stage directions
  in F always refer to 'Witches', although
  in the dialogue they are invariably the
  'Weïrd Sisters', see 1.3.32 and Introduction p. 3. The Fates are, in many myth-
  ologies, three, a number which has (like
  most prime numbers) many symbolic
  significances (e.g. the Trinity). How the
  Witches make their entry is not clarified,
  but see 1.3.78.
3 **hurly-burly** commotion, confusion—
  originally used of noble fighting, but in
  the sixteenth century commonly confined to the confusion of civil war
8 **Graymalkin** common name for a cat,

especially as a witch's familiar; possibly
pronounced, as it was sometimes spelt,
'Grimalkin'.
9–11 See Appendix A.
9 **Paddock** a toad, also a familiar
10 **Anon** immediately (the looser sense
  'soon' did also occur by Shakespeare's
  time)
11 **Fair is foul, and foul is fair** Sententious
  phrases such as 'fair without but foul
  within' were common (Dent F29; *Much
  Ado* 4.1.103), but the paradox 'foul is
  fair' was not: it becomes a *leitmotif* of the
  play (e.g. 1.3.38).

1.2   *Alarum within.*
       Enter King Duncan, Malcolm, Donalbain, Lennox,
       with Attendants, meeting a bleeding Captain
DUNCAN
What bloody man is that? He can report,
As seemeth by his plight, of the revolt
The newest state.
MALCOLM   This is the sergeant
Who like a good and hardy soldier fought
'Gainst my captivity.—Hail brave friend:
Say to the King the knowledge of the broil
As thou didst leave it.
CAPTAIN   Doubtful it stood,
As two spent swimmers that do cling together
And choke their art: the merciless Macdonald—
Worthy to be a rebel, for to that   10
The multiplying villainies of nature
Do swarm upon him—from the Western Isles
Of kerns and galloglasses is supplied,

1.2.0.1 *Duncan,*] *not in* F   13 galloglasses] F2; Gallowgrosses F1

1.2.0 **Alarum** A trumpet 'call to arms' her-
alding the approach of the army.
3 **sergeant** The title does not contradict
'*Captain*': a sergeant was originally a
tenant by military service, attending a
knight, and so equivalent to a commissioned officer in modern ranking (*OED* 3).
5 **'Gainst my captivity** i.e. against my being
taken prisoner
6 **knowledge** Used for information, intelligence (in the military sense) (*OED* II.12).
broil Used specifically for military engage-
ments when confused, especially in civil
war, whereas 'battle' was clearer both in
issue and in honour. *Othello* 1.3.87, 'feats
of broils and battle', is not merely tautologous.
7–42 The Captain's speech has two peculi-
arities: 1. that it uses the high-flown
rhetoric of the classical Messenger; 2.
that the Folio lineation is more than
usually erratic and cannot by any ingenuity be completely regularized. I assume
that Shakespeare was adapting a formal
device to the realistic circumstances of a
severely wounded man who eventually
loses coherence, but that the composi-

tor's bad habits have added irrelevant
irregularity. See Appendix A.
9 **choke their art** frustrate their skills as
swimmers (often, of course, by choking
each other)
10–12 The rhetorical phrasing is obscure.
Abbott (§ 186) stated that 'for to', later
used only with verbs, was used in Elizabethan English with nouns, and glossed
'to that end'; I prefer 'to that effect', i.e.
his natural rebelliousness breeds rebellion.
'Multiplying' and 'swarm' suggest a metaphor from insects, but it is ambiguous
whether it refers to Macdonald's self-
generating rebelliousness, or to his at-
tracting other rebels to himself—as in the
next two lines.
12 **Western Isles** Alternative name for the
Hebrides, off the West Coast of Scotland.
13 **kerns and galloglasses** F's 'Gallowgrosses' is presumably a misreading of an
unusual word. Kerns were foot-soldiers,
specifically associated with the 'wild
Irish' (confused with Scottish High-
landers and Islanders); galloglasses were
close attendants on Irish or Highland
chiefs.

And Fortune on his damnèd quarry smiling
Showed like a rebel's whore; but all's too weak,
For brave Macbeth—well he deserves that name—
Disdaining Fortune, with his brandished steel
Which smoked with bloody execution,
Like Valour's minion carved out his passage
Till he faced the slave—   20
Which ne'er shook hands nor bade farewell to him,
Till he unseamed him from the nave to th' chops,
And fixed his head upon our battlements.

DUNCAN

O valiant cousin, worthy gentleman.

CAPTAIN

As whence the sun 'gins his reflection
Shipwrecking storms and direful thunders break,
So from that spring, whence comfort seemed to come,
Discomfort swells. Mark, King of Scotland, mark,
No sooner justice had, with valour armed,
Compelled these skipping kerns to trust their heels,   30

---

14 quarry] F, OXFORD; quarrel HANMER   26 thunders break] POPE; Thunders F1; Thunders breaking F2

14–15 **Fortune ... whore** See the proverb 'Fortune is a strumpet' (Dent F603.1). Fortune is here personified as the Roman goddess Fortuna.
14 **his** may refer to Macdonald or to Fortune, since 'his' was often used for 'her'.
  quarry a common form of 'quarrel', i.e. Macdonald's rebellion; but it may also be that Fortune's smile is seducing his 'quarry', Duncan, who trusts Macdonald's loyalty. Holinshed refers to 'rebellious quarrel', but that does not justify emendation.
16 **name** good name, i.e. reputation as 'brave'
17 **steel** sword
18 **smoked** steamed, from the hot blood (cf. 'reeking' in l. 39)
19 **Valour's minion** Another personification; 'minion' is a loved favourite (from French 'mignon').
20 See Appendix A.
21 **Which** i.e. Macbeth
  shook hands Used of farewell as well as greeting: military courtesy did not necessarily consist only of courtesies before the fight.
22 **unseamed him** ripped him up
  nave centre, hence 'navel' (the words were confused); or, probably, 'crotch'
  chops jaws
24 **cousin** Used for any member of an extended family but also, as here, by a sovereign of a nobleman of his own court.
25 **his** its (as commonly)
  reflection literally 'turning back', i.e. at the equinoxes (see l. 27)
26 Hunter justifies F1, claiming that the rhythmic and syntactic discord is resolved by 'come' in l. 27; I find it impossible to deliver, and so accept Pope's variation on F2.
27 **spring** both the season (see l. 25) and the sun as 'source' of comfort (hence 'swells' in l. 28)
30 **skipping ... heels** 'skipping' depends on toes, but it also had the sense (as now) of absconding (OED skip v.¹, 2b)—hence they run away, as in 'take to their heels'.

But the Norwegian lord, surveying vantage,
With furbished arms and new supplies of men
Began a fresh assault.

DUNCAN

Dismayed not this our captains, Macbeth and Banquo?

CAPTAIN

Yes—as sparrows eagles, or the hare the lion:
If I say sooth, I must report they were
As cannons overcharged with double cracks,
So they doubly redoubled strokes upon the foe;
Except they meant to bathe in reeking wounds,
Or memorize another Golgotha,   40
I cannot tell—but I am faint,
My gashes cry for help.

DUNCAN

So well thy words become thee as thy wounds,
They smack of honour both. Go get him surgeons.

Enter Ross and Angus

Who comes here?

MALCOLM   The worthy Thane of Ross.

LENNOX

What a haste looks through his eyes! So should he
    look

31, 49 Norwegian] F (Norweyan)   32 furbished] F (furbusht)   46 a haste] F1; haste F2,
OXFORD

31 **Norwegian** first recorded in 1607; F's
'Norweyan' is found only in this play (see
also l. 49 and 1.3.95)
surveying vantage seeing a position
likely to give superiority; 'survey' implies
weighing up the whole position, and was
sometimes used simply for 'perceive'
(*OED* 4 and 4b)

32 **furbished** literally 'cleared of rust',
'brightened', but figuratively 'cleaned
up', 'revived'

32–5 See Appendix A.

36 **say sooth** speak the truth

37 **overcharged** overloaded
cracks the noise of a shot, here used for
the shot itself

38 The long line is no problem to speak (see
Appendix A). 'doubly' is a very likely
reduplication (however extra-metrical),
as in *Richard II* 1.3.80, especially here
with 'double' in l. 37, so I see no reason
to suppose a compositor's error. The

effect is onomatopoeic, as in Dryden's
'Song for St. Cecilia's Day, 1687',
ll. 19–30: 'The double double double
beat | Of the thundring drum.'

39 **Except** unless
reeking 'reek' was used of steam,
whether or not the exhalation was unpleasant; see 'smoked' in l. 18.

40 **memorize** cause to be remembered
Golgotha From Hebrew 'skull', 'charnel-
house', adopted into English from Gospel
accounts of the crucifixion at Golgotha,
'a place of a skull'; hence the allusion is
both impressive and ironic.

41–2 See Appendix A.

44 F has no direction for attention to the
Captain; possibly someone leaves to fetch
help, more probably a soldier helps the
Captain off stage.

45 **Thane** a baron, commonly a clan chief,
in Scotland

46 **a haste** Oxford comments that 'the in-

That seems to speak things strange.

ROSS   God save the King.

DUNCAN

Whence cam'st thou, worthy Thane?

ROSS   From Fife, great King,

Where the Norwegian banners flout the sky

And fan our people cold.   50

Norway himself, with terrible numbers,

Assisted by that most disloyal traitor,

The Thane of Cawdor, began a dismal conflict

Till that Bellona's bridegroom, lapped in proof,

Confronted him with self-comparisons,

Point against point, rebellious arm 'gainst arm,

Curbing his lavish spirit; and to conclude,

The victory fell on us—

DUNCAN

Great happiness!

ROSS   —that now Sweno,

The Norways' King, craves composition;   60

Nor would we deign him burial of his men

Till he disbursèd, at Saint Colum's Inch,

---

60 Norways'] STEEVENS 1778; Norwayes F1; Norway's F2   62 Colum's] F (*Colmes*)

definite article with "haste" was rare in Shakespeare, and "a" is the word most frequently interpolated.' F2's preferences are without authority, but it can be right; however, the text is rhythmically possible and effectively stresses 'haste'.

47 **seems to** appears to (an obsolete sense)

49 **flout** mock, insult

51 **Norway himself** i.e. the King of Norway

53 **dismal** ill-fated, evil; the word derived from the *dies mali* or *Aegyptiaci*, unlucky or evil days widely regarded in medieval Europe, and incorporated in 16th- and 17th-century calendars. Thomas, pp. 735–8, says that it was an essential feature that no one knew why they were unlucky nor why they were 'Egyptian' and there was no agreement on exact distribution (agricultural or astrological explanations were rationalizations). The Church condemned them as pagan, but some 'dismal days' were still avoided up to this century.

54 **Bellona's bridegroom** a rhetorical flourish to celebrate Macbeth's military glory

as his wedding to the Roman goddess of war

54 **lapped** wrapped, possibly with its erotic sense too, suggested by 'bridegroom' **proof** proved or tested power, originally of arms and armour (*OED* II.6)

55 **self-comparisons** i.e. matched him in every respect (ironically, in treachery as well)

56 **Point against point** 1. detail; 2. sword-point

57 **lavish** wild, prodigal

58–60 I assume that Ross completes his own verse line simultaneously with Duncan's extra-metrical exclamation (see Appendix A).

60 I suspect F's 'Norwayes' may be a misreading of 'Norweyan' (l. 31); as it stands 'composition' seems to me to need five syllables (which is possible). **composition** truce (*OED* III.23, treaty, *arch.*)

62 **Saint Colum's Inch** Inchcomb, St. Columba's island ('inch'), in the Firth of Forth; Shakespeare apparently wrote 'Colme' as a two-syllable word

Ten thousand dollars to our general use.

DUNCAN

No more that Thane of Cawdor shall deceive

Our bosom interest: go pronounce his present death,

And with his former title greet Macbeth.

ROSS I'll see it done.

DUNCAN

What he hath lost, noble Macbeth hath won.   *Exeunt*

I.3   *Thunder.*

Enter the three Witches

FIRST WITCH Where hast thou been, sister?

SECOND WITCH Killing swine.

THIRD WITCH Sister, where thou?

FIRST WITCH

A sailor's wife had chestnuts in her lap,

And munched, and munched, and munched. 'Give

me', quoth I;

'Aroynt thee, witch', the rump-fed ronyon cries.

Her husband's to Aleppo gone, master o'th' *Tiger*;

But in a sieve I'll thither sail,

---

63 **dollars** English form of Danish *rigsdalers*, in fact a sixteenth-century coinage; Holinshed only said 'a great sum of gold'.

65 **bosom interest** share or right in intimate trust

1.3.6 **Aroynt ... cries** The general sense seems to be '"Be off, witch" the overfed harridan cries', but the actual words are elusive: 'aroynt', 'rump-fed', and 'ronyon' are all peculiar to Shakespeare; 'aroynt thee, witch' is also in *Lear* 3.4.117, 'ronyon' in *Merry Wives* 4.2.172, also of a witch. Very likely they all existed in slang but did not get into print elsewhere. The First Witch's anger is not simply at being refused chestnuts, but also at the insulting term 'witch'—they speak of themselves as 'Weïrd Sisters' (l. 32; see Introduction, pp. 2–4). She seems to be returning the compliment with interest: 'rump-fed' may mean 'fat-buttocked' as well as 'fed on rump' (offal as well as good meat); 'ronyon' may derive from French 'rognon' 'kidney' (originally human or animal); it had archaically a slang sense of

the testicles of certain animals (E. Littré, *Dictionnaire de la langue française*, vi; Paris, 1958). 'Rogne' is still a feminine noun for mange and similar diseases, derived from Walloon 'ragn'; Professor Irène Simon of Liège called my attention to J. Haust, *Dictionnaire Liègeois* (Liège, 1933), which gives the phrase 'mâle rogne' for 'méchante femme', 'evil woman'. I take it these two words coalesced and crossed the channel as a term of abuse; it is still common for the genitals of one sex to be used for abuse of either (e.g. 'balls', 'cunt'); the First Witch adds a suggestion of buggery in 'rump-fed ronyon'. (*OED* prefers 'aroint' and 'runnion', but the changes seem pointless.)

7 *Tiger* A common name for ships; its use here may refer to (big) cats as witches' familiars. Two 'Tigers' are discussed in Introduction, pp. 61–2.

8 **sieve** It was common in the mythology of witches, and characteristic of their inverted powers, to sail in a bottomless boat.

And like a rat without a tail,
    I'll do, I'll do, and I'll do.   10
SECOND WITCH I'll give thee a wind.
FIRST WITCH Thou'rt kind.
THIRD WITCH And I another.
FIRST WITCH
    I myself have all the other,
    And the very ports they blow,
    All the quarters that they know
    I'th' shipman's card.
    I'll drain him dry as hay;
    Sleep shall neither night nor day
    Hang upon his penthouse lid;   20
    He shall live a man forbid.
    Weary sev'n-nights, nine times nine,
    Shall he dwindle, peak, and pine;
    Though his bark cannot be lost,
    Yet it shall be tempest-tossed.
    Look what I have.
SECOND WITCH Show me, show me.
FIRST WITCH
    Here I have a pilot's thumb,
    Wrecked as homeward he did come.
      Drum within
THIRD WITCH
    A drum, a drum: 30
    Macbeth doth come.

---

9 **rat without a tail** Steevens (1793) stated that old writers rationalized defective transformation by claiming that no part of a woman could be changed to a tail, but he cites no authority. Thomas (p. 529) states that allusions to animal transformation were rare.

11 **give thee a wind** Thomas does not mention this idea; Muir cites fictional instances from Nashe, Fletcher, and Drayton.

12 **kind** In the original sense of 'natural', because witches normally sold winds, as well as in the modern sense.

14 **other** Singular was often used for plural.

15 **they blow** i.e. when they blow; possibly 'blow from', but onshore winds can equally prevent entry to a port

16 **quarters** the four principal points of the compass

17 **card** compass-card

20 **penthouse lid** literally, sloping roof of a lean-to building, used figuratively of eye-coverings; in *LLL* 3.1.17, it is a hat, though more commonly, as here, eyelids or brows.

21 **forbid** laid under a ban (i.e. cursed) (*OED* 2f)

23 **peak, and pine** Both verbs mean 'waste'—a common phrase.

ALL

> The Weïrd Sisters, hand in hand,
> Posters of the sea and land,
> Thus do go, about, about,
> Thrice to thine, and thrice to mine,
> And thrice again, to make up nine.
> Peace, the charm's wound up.
>> *Enter Macbeth and Banquo*

MACBETH

> So foul and fair a day I have not seen.

BANQUO

> How far is't called to Forres?—What are these,
> So withered, and so wild in their attire,   40
> That look not like th'inhabitants o'th' earth
> And yet are on't?—Live you, or are you aught
> That man may question? You seem to understand me,
> By each at once her choppy finger laying
> Upon her skinny lips. You should be women,
> And yet your beards forbid me to interpret
> That you are so.

MACBETH (*to the Witches*)

>   Speak if you can: what are you?

1.3.32 Weïrd] F (weyward)   39 Forres] POPE; Soris F

32 **Weïrd Sisters** Modern spellings are misleading: Compositor A prints 'weyward' (and in 1.5.7 and 2.1.21), where Compositor B prints 'weyard' (3.1.2, 3.4.134, 4.1.151) which is a clearer spelling of 'weïrd' as a two-syllable word, distinct from the modern 'weird'. The Weïrd were the Fates in Anglo-Saxon (and later) mythology, responsible, like the classical Parcae for destiny, and therefore divination, rather than for witchcraft. See Introduction, pp. 2–4, 78–9.

33 **Posters** travellers who are swift (as in 'post-haste')

35–6 The implied ritual might divide the phrases between the witches.

35 **Thrice** Used as a magic number, here related to the trinity of the Fates as in classical mythology.

37 **wound up** Used figuratively for setting anything in readiness, especially where tension is involved (it was already applied to watch-springs by 1601) (*OED* f, and

'wind', *v.* ¹ 17).

39 **called** said to be. *OED* does not have this sense, but it has survived in the USA (Webster's *New World Dictionary* (1957), *v.* 10).

39–47 The peculiarity of these lines is that they describe so elaborately, though deliberately vaguely, figures which the audience can actually see. No wonder designers vary enormously in how to make them up and clothe (or unclothe) them, and as to whether they resemble each other or not. It is at least possible that they do not look like what Banquo describes. See Introduction, pp. 2–3.

43 **question** interrogate. See *Hamlet* 1.4.24–5, 'Thou com'st in such a questionable shape | That I will speak to thee.'

44 **choppy** chapped, with skin cracked

46 **beards** Thomas says 'it was proverbial that bearded women were likely to be witches'; Shakespeare mentions it in *Merry Wives* 4.2.170–2.

FIRST WITCH
 All hail Macbeth, hail to thee Thane of Glamis.

SECOND WITCH
 All hail Macbeth, hail to thee Thane of Cawdor.

THIRD WITCH
 All hail Macbeth, that shalt be King hereafter.   50

BANQUO
 Good sir, why do you start, and seem to fear
 Things that do sound so fair?—I'th' name of truth,
 Are ye fantastical, or that indeed
 Which outwardly ye show? My noble partner
 You greet with present grace, and great prediction
 Of noble having, and of royal hope,
 That he seems rapt withal; to me you speak not.
 If you can look into the seeds of time,
 And say which grain will grow, and which will not,
 Speak then to me, who neither beg nor fear   60
 Your favours, nor your hate.

FIRST WITCH Hail.

SECOND WITCH Hail.

THIRD WITCH Hail.

FIRST WITCH
 Lesser than Macbeth, and greater.

SECOND WITCH
 Not so happy, yet much happier.

THIRD WITCH
 Thou shalt get kings, though thou be none:
 So all hail Macbeth, and Banquo.

FIRST WITCH
 Banquo, and Macbeth, all hail.

MACBETH
 Stay, you imperfect speakers, tell me more: 70
 By Sinell's death I know I am Thane of Glamis,
 But how, of Cawdor? The Thane of Cawdor lives

---

53 **fantastical** in the obsolete sense 'existing only in the imagination'
55 **present grace** i.e. present title **prediction** foretelling
56 i.e. of nobility (Cawdor) already possessed, and hope of royalty to come
57 **rapt** carried away in spirit
67 **get** beget
70 **imperfect** incomplete (but with a secondary sense of 'evil')
71 **Sinell** Holinshed's name for Macbeth's father.

A prosperous gentleman; and to be King
Stands not within the prospect of belief,
No more than to be Cawdor. Say from whence
You owe this strange intelligence, or why
Upon this blasted heath you stop our way
With such prophetic greeting? Speak, I charge you.

Witches vanish

BANQUO
The earth hath bubbles, as the water has,
And these are of them; whither are they vanished?   80

MACBETH
Into the air; and what seemed corporal melted
As breath into the wind. Would they had stayed.

BANQUO
Were such things here as we do speak about?
Or have we eaten on the insane root
That takes the reason prisoner?

MACBETH
Your children shall be kings.

BANQUO   You shall be King.

MACBETH
And Thane of Cawdor too: went it not so?

BANQUO
To th' self-same tune, and words.—Who's here?

Enter Ross and Angus

ROSS
The King hath happily received, Macbeth,
The news of thy success; and when he reads   90
Thy personal venture in the rebels' fight,

---

73 **prosperous** flourishing
74 **prospect** forward vision, possible expectation
78.1 ***Witches vanish*** A disappointingly vague direction: evidently an ingenious illusion was called for, or else ll. 79–82 describe what the audience must know as different from what they saw (see note on ll. 39–47). The resources of the Globe Theatre included smoke, trapdoors, and at least one winch (but it is unlikely that three actors could have been flown fast enough).
81 **corporal** bodily, corporeal

84 **on** commonly used for 'of'
insane root An attributive use of 'insane', i.e. a root which causes insanity, but also playing on the 'root of all insanity', i.e. fundamental unreason, chaos. A number of roots are powerful narcotics, e.g. henbane, hemlock, mandrake.
88 **To th' self-same tune, and words** Banquo passes it off by a frivolous play on 'went', as in 'how does the song go'.
90 **reads** considers (the original sense of the word)
91 **venture** danger, enterprise

His wonders and his praises do contend
Which should be thine, or his. Silenced with that,
In viewing o'er the rest o'th' self-same day,
He finds thee in the stout Norwegian ranks,
Nothing afeard of what thyself didst make
Strange images of death. As thick as hail
Came post with post, and every one did bear
Thy praises in his kingdom's great defence
And poured them down before him.

ANGUS   We are sent   100
To give thee from our royal master thanks,
Only to herald thee into his sight,
Not pay thee.

ROSS

And for an earnest of a greater honour
He bade me, from him, call thee Thane of Cawdor:
In which addition, hail most worthy Thane,
For it is thine.

BANQUO   What, can the devil speak true?

MACBETH

The Thane of Cawdor lives; why do you dress me
In borrowed robes?

ANGUS   Who was the Thane, lives yet,

---

96 make ∧] F; make, ROWE, OXFORD   97–8 hail | Came] ROWE; Tale | Can F

92–3 Thoroughly confusing lines which enact the confusion of mind that silences
Duncan, to say nothing of the confusion
of battle which Ross goes on to describe.
Oddly enough, it is syntactically correct
that Duncan can with equal propriety
speak of 'his wonder and his praise' or of
'thy wonders and thy praise', but hard to
see why that should reduce him to silence. The lines do still function as hyperbolic amazement.
93 **that** i.e. that state of affairs
96–7 **Nothing...death** 1. not afraid of those
whom you proceeded to kill; 2. not afraid
of the strange images of death which you
yourself had made (by killing them all);
W. Empson, *Seven Types of Ambiguity*
(1930, 3rd edn., 1953), 45, rightly insists that the residual force here is of
'Strange images of death' which later *do*
terrify Macbeth after he has murdered
Duncan. Adding a comma after 'make'
seriously reduces the ambiguity.

97–8 **hail | Came** Rowe's emendation is
generally accepted, and goes well with
'poured' in l. 100; F's 'Tale | Can' is just
about interpretable, but could hardly be
spoken intelligibly. For the proverbial 'as
thick as hail' see Dent H11.
103 **Not pay thee** i.e. not to bring adequate
reward
104 **earnest** pledge of what is to come
greater honour Literally we hear of no
greater honour from Duncan, unless it is
his visit to Macbeth's castle; but the
vague rhetorical phrase has obvious
ironic reference to the Witch's promise of
kingship.
106 **addition** commonly used of titles, an
addition to a person's name and rank
107 **What ... true** 'The Devil sometimes
speaks the truth' was certainly proverbial
later (Dent D266), but it is not recorded
before Shakespeare, *Richard III* 1.2.73, 'O
wonderful, when devils tell the truth!'
See also ll. 124–7 below.

But under heavy judgement bears that life   110
Which he deserves to lose.
Whether he was combined with those of Norway,
Or did line the rebel with hidden help
And vantage; or that with both he laboured
In his country's wrack, I know not.
But treasons capital, confessed and proved,
Have overthrown him.
MACBETH (*aside*)   Glamis, and Thane of Cawdor:
The greatest is behind.—Thanks for your pains.
(*To Banquo*) Do you not hope your children shall be
    kings,
When those that gave the Thane of Cawdor to me,   120
Promised no less to them?
BANQUO   That trusted home
Might yet enkindle you unto the crown,
Besides the Thane of Cawdor. But 'tis strange;
And oftentimes, to win us to our harm,
The instruments of darkness tell us truths,
Win us with honest trifles, to betray's
In deepest consequence.—
Cousins, a word, I pray you.
MACBETH (*aside*)   Two truths are told
As happy prologues to the swelling act

111–15 See Appendix A.
113 **line** reinforce (figurative use from lining
    clothes)
114 **vantage** benefit
115 **wrack** ruin (associated with ven-
    geance); the word derives from OE,
    meaning vengeance, but became con-
    fused with 'wreck'
116 **treasons capital** i.e. capital treasons
118 **The greatest is behind** Proverbially
    either best or worst was behind (Dent
    B318, W918).
120, 123 **Thane** Apparently in the sense of
    'thanedom', not in *OED*; but its use here
    derives from the title, and may mean no
    more.
121 **home** (adverb) thoroughly
122 **enkindle** literally, cause to blaze up;
    figuratively, cause passionate excite-
    ment, violence, etc. It is stronger, there-
    fore, than 'cause you to hope for' (Brad-
    ley).

124–7 See l. 107. A traditional charge
    against devilish appropriation of oracular
    prophecy; Muir quotes James I, *Demon-
    ology*, in *Works*, 1616, p. 98: 'for that old
    and crafty serpent being a spirit, he easily
    spies our affections, and so conforms
    himself thereto to deceive us to our
    wrack'.
127–8 Macbeth's half-line completes Ban-
    quo's 'In deepest consequence', ignoring
    'Cousins, a word, I pray you'; for the rest
    of the scene (until l. 154) Banquo and
    Macbeth do not directly communicate
    but their speeches are related in rhythm
    as in thought.
129 **prologues ... act** as in a play
    swelling figuratively, magnifying, exalt-
    ing (*OED* 5d)—i.e. as the act develops and
    expands. *OED* gives the musical sense of
    'swell' (crescendo) as later (1749), but it
    is strikingly apt here with 'theme', since
    the whole phrase is a superb crescendo.

Of the imperial theme.—I thank you gentlemen—   130
This supernatural soliciting
Cannot be ill, cannot be good. If ill,
Why hath it given me earnest of success,
Commencing in a truth? I am Thane of Cawdor.
If good, why do I yield to that suggestion
Whose horrid image doth unfix my hair
And make my seated heart knock at my ribs
Against the use of nature? Present fears
Are less than horrible imaginings:
My thought, whose murder yet is but fantastical,   140
Shakes so my single state of man, that function
Is smothered in surmise, and nothing is
But what is not.
BANQUO   Look how our partner's rapt.
MACBETH (*aside*)
If chance will have me King, why chance may crown
    me,
Without my stir.
BANQUO   New honours come upon him
Like our strange garments cleave not to their mould
But with the aid of use.
MACBETH (*aside*)   Come what come may,
Time and the hour runs through the roughest day.

130 **theme** 1. topic; 2. principal melody in contrapuntal music
131 **soliciting** inciting, alluring
133 **earnest** See l. 104.
136 **horrid** standing on end (hair); hideous, horrifying
unfix unfasten, unsettle
137 **seated** having its settled position in the body (*OED*, Seat *v.*⁵ 5c)
138 **use** usage, customary arrangement
140 **whose murder** i.e. the thought of murder, which has been hinted with gathering intensity throughout the scene but only here comes to explicit recognition
141 **single state of man** 'single' had the sense (*OED* 4) of 'undivided, unbroken, absolute'; here it surely means 'unified', and I agree with Muir that it alludes to the microcosm, the body, spirits, etc., unified by an undivided soul. Macbeth is threatened with self-division and disintegration: the nearest modern equivalent would be 'integrity'.

141 **function** the special kind of action proper to anything physical or mental (*OED* 3); here, the proper functioning of a man whose sane mind governs his action.
142 **surmise** unwarranted conjecture, hence imagination (*OED* 5)
142–3 **nothing is | But what is not** The displacement of actuality by illusion, a key phrase for the play both in its staging and its language (see Introduction, pp. 1–34).
143 **rapt** See l. 57.
145–7 i.e. the honours are uncomfortable until familiar, like new clothes adjusting to one's body
147–8 **Come what ... day** Combining two sententious phrases, 'Come what come may' and 'The longest day has an end', Dent C529, D20.
148 Time carries on despite the roughest of days, and whatever will happen does happen.

BANQUO

Worthy Macbeth, we stay upon your leisure.

MACBETH

Give me your favour: my dull brain was wrought   150
With things forgotten. Kind gentlemen, your pains
Are registered where every day I turn
The leaf to read them. Let us toward the King.
(*To Banquo*) Think upon what hath chanced, and at
          more time,
The interim having weighed it, let us speak
Our free hearts each to other.

BANQUO   Very gladly.

MACBETH

Till then enough.—Come friends.   *Exeunt*

I.4   *Flourish.*
          Enter King Duncan, Lennox, Malcolm, Donalbain,
          and Attendants

DUNCAN

Is execution done on Cawdor? Or not
Those in commission yet returned?

MALCOLM   My liege,

They are not yet come back. But I have spoke
With one that saw him die, who did report
That very frankly he confessed his treasons,
Implored your highness' pardon, and set forth
A deep repentance. Nothing in his life
Became him like the leaving it. He died
As one that had been studied in his death

1.4.0.1 *Duncan*] not in F   1 Or] F1; Are F2, OXFORD

150 **wrought** The old past participle of
'work', here in the sense of 'strained',
'disturbed' (as in 'overwrought'); see
OED, 'work' *v*, II.10, but more specifically for the *sb.* 6.
151 **forgotten** an excuse, the opposite of the
truth
152–3 **registered … read them** i.e. noted in
a journal whose pages I turn daily
155 **The interim having weighed it** i.e. 'give
ourselves time to weigh what has
chanced'; or, perhaps, 'having weighed

(considered) what has chanced in the
mean time'
1.4.1 **Or not** i.e. or are not. The visual
similarity of 'Or' and 'Are' may support
the common emendation; alternatively,
the MS may have read 'Or are'; but the
phrase is intelligible as it stands.
2 **in commission** given the commission to
carry out the execution
9 **studied** carefully prepared, as an actor
'studies' his role, i.e. learns his lines and
prepares his performance

To throw away the dearest thing he owed,   10
As 'twere a careless trifle.

DUNCAN   There's no art
To find the mind's construction in the face;
He was a gentleman on whom I built
An absolute trust.

Enter Macbeth, Banquo, Ross, and Angus
                         O worthiest cousin,
The sin of my ingratitude even now
Was heavy on me. Thou art so far before
That swiftest wing of recompense is slow
To overtake thee. Would thou hadst less deserved,
That the proportion both of thanks, and payment,
Might have been mine; only I have left to say,   20
More is thy due, than more than all can pay.

MACBETH
The service and the loyalty I owe,
In doing it, pays itself.
Your highness' part, is to receive our duties;
And our duties are to your throne and state,
Children and servants, which do but what they should
By doing everything safe toward your love
And honour.

DUNCAN   Welcome hither:
I have begun to plant thee, and will labour
To make thee full of growing.—Noble Banquo,   30
That hast no less deserved, nor must be known
No less to have done so—let me enfold thee,
And hold thee to my heart.

BANQUO   There if I grow,

---

10 **owed** 1. owned; 2. owed, as a life is 'owed
  to God'
11–12 Based on the proverb 'the face is no
  index to the mind' Dent F1.1.
16–18 The meaning is clear enough, the
  only strangeness is the rhetorical amplification which is characteristic of this
  scene in contrast to the compression of
  Macbeth's language in 1.3.
19–20 **That the proportion ... mine** i.e. that
  I could have made thanks and payment
  proportional to your deserts

20 **only I have left to say** i.e. I have left to say
  only that
22–3 A ceremonial amplification of the
  proverbial 'virtue is its own reward',
  Dent V81.
23–8 See Appendix A.
27 **safe toward** i.e. to secure your safety and
  so merit ...
31–2 **no less ... nor ... no less** Multiple
  negatives did not necessarily cancel each
  other; modern English would use 'and'
  for 'nor' here.

The harvest is your own.

DUNCAN   My plenteous joys,
Wanton in fullness, seek to hide themselves
In drops of sorrow.—Sons, kinsmen, thanes,
And you whose places are the nearest, know
We will establish our estate upon
Our eldest, Malcolm, whom we name hereafter
The Prince of Cumberland; which honour must   40
Not unaccompanied invest him only,
But signs of nobleness like stars shall shine
On all deservers. From hence to Inverness,
And bind us further to you.

MACBETH
The rest is labour, which is not used for you;
I'll be myself the harbinger, and make joyful
The hearing of my wife with your approach.
So humbly take my leave.

DUNCAN   My worthy Cawdor.

MACBETH (*aside*)
The Prince of Cumberland: that is a step
On which I must fall down, or else o'er-leap,   50
For in my way it lies. Stars hide your fires,
Let not light see my black and deep desires,
The eye wink at the hand—yet let that be
Which the eye fears, when it is done, to see.   *Exit*

DUNCAN
True, worthy Banquo; he is full so valiant,

34–6 i.e. I weep for joy
35 **Wanton** extravagant, unrestrained
37 **nearest** nearest to the throne (cf. 'next-of-kin')
38 **establish our estate** settle or confirm our position, property, etc.—make Malcolm his official heir to the throne, etc.; Scottish monarchy did not necessarily follow primogeniture (Introduction, p. 68)
40 **Prince of Cumberland** The title used for the Scottish heir to the throne, as Wales is for the English; Cumberland was held by the Scots under the English crown.
45 A convoluted courtesy, responding to Duncan's 'bind' implying that he's laying a burden on Macbeth by visiting his castle; the apparent meaning is 'resting would be a burden compared to serving you', but 'rest' and 'labour' interchange confusingly in 'which'. The line can be read in quite another sense as 'what remains is labour [preparing for your visit], which you are not accustomed to' (Foakes); my reading was suggested by the general editor, and is supported by Hunter. The exchange is very similar to 1.6.11–15.
46 **harbinger** one sent to purvey for lodgings for army or royalty (*OED* 2)
53 **The eye wink at the hand** i.e. let the eye not see what the hand is doing
55 Clearly Duncan and Banquo have been in conversation during Macbeth's aside and exit.
full fully, entirely (the adverbial use was common in middle English)

And in his commendations I am fed,
It is a banquet to me. Let's after him,
Whose care is gone before to bid us welcome;
It is a peerless kinsman.　　　　*Flourish. Exeunt*

1.5　　*Enter Macbeth's wife alone, with a letter*

LADY MACBETH 'They met me in the day of success; and I
have learned by the perfectest report, they have more in
them than mortal knowledge. When I burned in desire
to question them further, they made themselves air, into
which they vanished. Whiles I stood rapt in the wonder
of it, came missives from the King, who all-hailed me
Thane of Cawdor, by which title, before, these Weïrd
Sisters saluted me, and referred me to the coming-on of
time with "Hail King that shalt be." This have I thought
good to deliver thee, my dearest partner of greatness,　　10
that thou mightst not lose the dues of rejoicing by being
ignorant of what greatness is promised thee. Lay it to
thy heart, and farewell.'
Glamis thou art, and Cawdor, and shalt be
What thou art promised; yet do I fear thy nature,
It is too full o'th' milk of human kindness
To catch the nearest way. Thou wouldst be great,
Art not without ambition, but without
The illness should attend it. What thou wouldst highly,
That wouldst thou holily; wouldst not play false,　　　20
And yet wouldst wrongly win. Thou'dst have, great
　　Glamis,
That which cries, 'Thus thou must do' if thou have it;

1.5.7 Weïrd] F (weyward)　　22 cries, 'Thus ... do'] F (cryes, | Thus thou must doe,)

56 **commendations** things that recommend
1.5.2 **perfectest** Compare 'imperfect speak-
　　ers' in 1.3.70.
　6 **missives** Usually 'letters', used speci-
　　fically of orders from a king; it was also
　　occasionally used, as here, of the messen-
　　ger, hence 'who all-hailed ...'
16 **kindness** 'kind' originally meant 'nat-
　　ural' and developed its modern sense
　　from 'natural goodness'; both are in-
　　volved here, for Lady Macbeth is

consciously invoking the unnatural (see
l. 40).
17 **nearest** most expeditious
19 **illness** badness
22 **'Thus thou must do'** Editors have vari-
　　ously expanded the inverted commas as
　　far as the end of the line, or to 'undone' in
　　l. 24; but F's commas support my read-
　　ing, and as Muir argued, this arrange-
　　ment makes best sense.
　　if thou have it i.e. if you are to have it

And that which rather thou dost fear to do,
Than wishest should be undone. Hie thee hither,
That I may pour my spirits in thine ear,
And chastise with the valour of my tongue
All that impedes thee from the golden round,
Which Fate and metaphysical aid doth seem
To have thee crowned withal.
> *Enter Messenger*

                        What is your tidings?

MESSENGER
The King comes here tonight.

LADY MACBETH   Thou'rt mad to say it.   30
Is not thy master with him? Who, were't so,
Would have informed for preparation.

MESSENGER
So please you, it is true; our Thane is coming;
One of my fellows had the speed of him,
Who almost dead for breath, had scarcely more
Than would make up his message.

LADY MACBETH   Give him tending,
He brings great news.   *Exit Messenger*
              The raven himself is hoarse
That croaks the fatal entrance of Duncan
Under my battlements. Come, you spirits

23–4 The meaning is clear, but the clotted syntax seems to obscure it, in close anticipation of Macbeth's language in 1.7.1–7.

25 **spirits** 1. immaterial qualities (transmitted by words); 2. courage, mettle; 3. distilled poison

26 **chastise** The stress is on the first syllable.

27 **golden round** crown

28 **metaphysical** supernatural

32 **informed** reported (used intransitively)

34 **had the speed of** outpaced

36 **Give him tending** take care of him

37–53 The whole speech has a density of language and imagery that makes commentary extremely difficult; it picks up from 1.3.128–43, and in several respects anticipates 1.7.1–28—see Introduction, pp. 14–16. The actor playing Lady Macbeth has to choose between enacting a ritual appeal to spirits literally invoked, or using the whole passage as the language of autosuggestion to bolster her own morale. Sarah Siddons changed her performance from the second to the first; the text suggests both, but it is hard to see how a performance can avoid choosing between them. (See Introduction, pp. 14–15.)

37 **raven** bird of omen, whose raucous cry betokens death; 'hoarse' suggests it is here more raucous and ominous even than usual. Ominous allusions to the raven were common, e.g. 'the croaking raven bodes death', Dent R33.

39–40 **spirits ... thoughts** spirits believed to wait (like devils) for the invitation of an evil thought to take possession of a human mind, and turn the thought into action

That tend on mortal thoughts, unsex me here,   40
And fill me from the crown to the toe, top-full
Of direst cruelty. Make thick my blood,
Stop up th'access and passage to remorse,
That no compunctious visitings of nature
Shake my fell purpose, nor keep peace between
Th'effect and it. Come to my woman's breasts
And take my milk for gall, you murd'ring ministers,
Wherever, in your sightless substances,
You wait on nature's mischief. Come, thick night,
And pall thee in the dunnest smoke of Hell,   50
That my keen knife see not the wound it makes,
Nor Heaven peep through the blanket of the dark
To cry, 'Hold, hold'.
     Enter Macbeth
               Great Glamis, worthy Cawdor,
Greater than both, by the all-hail hereafter,
Thy letters have transported me beyond

46 it] F (hit)

40 **mortal thoughts** 1. human thoughts; 2.
    deadly thoughts
    unsex me here 'sex' governs all the range
    of human experience ('mortal thoughts')
    that follows: kindness, remorse, pity,
    fertility; it is, in short, her humanity and
    not simply her feminity that Lady Mac-
    beth wishes away—the thoughts are
    very close to those which disturb Mac-
    beth in 1.7
42 **Make thick my blood** Healthy blood is
    clear, permitting the passage of 'natural
    spirits' (in this case pity and fear) to the
    brain; 'thick' blood obstructs their pas-
    sage.
44 **compunctious** pricking of conscience or
    heart
45 **fell** ruthless, cruel, etc.
45–6 **keep ... it** i.e. interpose peace between
    my purpose ('it') and its dire effect
47 **take my milk for gall** substitute gall for
    milk
48 **sightless** 1. blind; 2. invisible; 3. un-
    sightly, ugly (see 1.7.23)
    substances Spirits, like angels, had to
    take on 'substance', i.e. matter, in order
    to contact the material world, even

though not composed of the four elements which make up all sub-lunar
bodies; the visible form would be an
expression of its spiritual essence.
49 **wait on nature's mischief** lie in wait for
    disturbances of nature, which they can
    promote as they can respond to her evil
    thoughts because they are 'unnatural'
50 **pall** 1. cover, as with a cloak; 2. appal,
    become appalling
    dunnest dingiest brown/greyest, used for
    'murkiest'
51 **keen** 1. sharp; 2. eager—the knife's
    eagerness might be blunted if it saw the
    horror of the wound it will make
52 **blanket** Neo-classical insistence on ele-
    vated diction caused Johnson and even
    Coleridge to object to this. In fact, the
    vocabulary of this speech is as insistently
    literal as its *metaphoric* reference is ima-
    ginative.
54 **all-hail** Intensive of 'hail'; the witches
    reserved it for kingship (1.3.68–9) which
    is Lady Macbeth's thought here.
55 **letters have** The plural was used with
    singular meaning, as in Latin *litterae*
    (*OED* II.2b).

This ignorant present, and I feel now
The future in the instant.
MACBETH   My dearest love,
Duncan comes here tonight.
LADY MACBETH   And when goes hence?
MACBETH
Tomorrow, as he purposes.
LADY MACBETH   O never
Shall sun that morrow see.   60
Your face, my Thane, is as a book where men
May read strange matters; to beguile the time,
Look like the time, bear welcome in your eye,
Your hand, your tongue—look like th'innocent flower,
But be the serpent under't. He that's coming
Must be provided for; and you shall put
This night's great business into my dispatch,
Which shall to all our nights and days to come
Give solely sovereign sway and masterdom.

MACBETH
We will speak further.
LADY MACBETH   Only look up clear;   70
To alter favour, ever is to fear.
Leave all the rest to me.   *Exeunt*

---

62 time,] THEOBALD; time. F

56 **ignorant** unknowing. The point is that
the present always makes the future
invisible, as in *Winter's Tale* 1.2.396-7,
'imprison't not | In ignorant conceal-
ment'.
60 The short line implies a significant pause:
Mrs Siddons used it to stare into Macbeth's eyes.
62 **beguile the time** deceive the world at the
present time, as in the proverb 'to beguile
the time with a fair face', Dent T340.1
(see 1.7.82)
64-5 **look ... under't** Muir attributes this
idea to Virgil's *Eclogues* iii. 93; there, as
elsewhere in Shakespeare, it describes
snakes as such, but here the colloca-
tion of 'innocent' and 'serpent' strongly
suggests an allusion to Satan; it is
not in Genesis nor in the usual elabora-
tions of Christian mythology, and

though Milton makes the connection
in *Paradise Lost*, ix. 908-12, it is not
the actual way that Satan deceives Eve.
'A snake in the grass' was common
proverbially (Dent S585), here elaborating 'fair face foul heart' (T340.1);
see 1.7.82-3.
67 **dispatch** 1. riddance; 2. speed. The word
was explicitly used of killing in the sixteenth century but had also clerical associations so it may imply 'management' as
well as 'murder'.
69 **solely** only, exclusively—hence, absolutely
70 **clear** unclouded (by guilt or fear)
71 The sentence is inverted to complete the
couplet by opposing 'fear' to 'clear'; the
sense is 'fear is always apt to alter one's
countenance (favour)'. See 1.7.83 and
3.2.30.

1.6 *Enter King Duncan, Malcolm, Donalbain, Banquo,*
Lennox, Macduff, Ross, Angus, and Attendants

DUNCAN

This castle hath a pleasant seat, the air
Nimbly and sweetly recommends itself
Unto our gentle senses.

BANQUO   This guest of summer,
The temple-haunting martlet, does approve,
By his loved mansionry, that the Heaven's breath
Smells wooingly here; no jutty, frieze,
Buttress, nor coign of vantage, but this bird
Hath made his pendant bed and procreant cradle;

1.6.0 *Enter King Duncan*] WILSON; *Hoboyes, and Torches. Enter King* F   4 martlet] ROWE; Barlet F
5 mansionry] THEOBALD; Mansonry F; masonry POPE 1728   6 jutty, frieze] STEEVENS, Iutty
frieze F; jutting frieze POPE

1.6.0.1 F calls for 'Hautboys and torches'
(i.e. musicians and torchbearers) for both
this scene and the next, and that may be
right; but Wilson pointed out both the
oddity of oboes for a scene emphatically
located out-of-doors, and of torches
where light is stressed in contrast to the
prevailing darkness. Muir's suggestion of
sunset seems to me an irrelevant
rationalization since there is no verbal
reference to it. Stage directions in F are
commonly added from prompter's notes
and since 1.6 is very short, this may well
have originated as a reminder to call the
openers for 1.7, where they are essential
to indicate darkness in an open-air
theatre (see Introduction, pp. 1–2).

1 **seat** situation (often used of large houses
themselves)

2 **Nimbly** quickly and lightly

3 **gentle** originally 'noble' in birth and
therefore in nature; here that is combined with the later sense which is still
in use

4 **temple-haunting martlet** house-martin
(or swallow); F's 'Barlet' may be an error
for 'marlet' which *OED* gives as a rare
obsolete word derived from OF 'merlette',
diminutive of 'merle' (blackbird), and
glosses 'martin or martlet'; 'martlet',
according to *OED*, is strictly the swift, but
in fact it was frequently used for swallow
or house-martin. They in fact nest on any
building, but Brathwait, *Survey of His-*
tory (1638), wrote 'the martin will not
build but in fair houses'. I assume that
'martlet' displaced 'marlet' and was
probably understood of any members of
the swallow family, of which house-
martins and swallows both habitually
nest under ledges on buildings. The
specific claim for temples sounds proverbial but is not in Tilley or Dent, and
may be an embellishment here to stress
the sanctity in the whole passage.

approve attest

5 **his loved mansionry** probably a play on
'the buildings he loves' and 'his loving
building of his nest' (i.e. masonry). The
word 'mansionry' is rare and *OED* is
unsure of its meaning.

6 **jutty** alternative form of 'jetty', meaning
a projecting part of a building, especially
an overhanging upper storey.

frieze the area between the lintel of a
door, window, etc., and the projecting
cornice above, not necessarily orna-
mented

7 **coign of vantage** strictly, a projecting
corner; but *OED* has this phrase as
common for any advantageous position

8 **pendant** As a noun, a triangular structure between roof and beam (architectural term, *OED* 6); this precisely describes
a martin's nest which does not simply
hang, the common sense of 'pendant';
but it is clearly adjectival here, 'pendant-
like'.

bed nest (*OED*, bed *sb.* I.2b, has the
figurative sense 'place of conjugal union,
and of procreation and childbirth')

Where they must breed and haunt, I have observed
The air is delicate.   10
    Enter Lady Macbeth

DUNCAN
  See, see, our honoured hostess:—the love
  That follows us sometime is our trouble,
  Which still we thank as love. Herein I teach you
  How you shall bid God 'ield us for your pains,
  And thank us for your trouble.

LADY MACBETH   All our service,
  In every point twice done, and then done double,
  Were poor and single business, to contend
  Against those honours deep and broad wherewith
  Your majesty loads our house; for those of old,
  And the late dignities heaped up to them,   20
  We rest your hermits.

DUNCAN   Where's the Thane of Cawdor?
  We coursed him at the heels, and had a purpose
  To be his purveyor; but he rides well,
  And his great love, sharp as his spur, hath holp him
  To his home before us. Fair and noble hostess,
  We are your guest tonight.

9 must] F; most ROWE   10.1 *Macbeth*] *not in* F

9 **must** 1. are resolved to (*OED* 3), normally used with the first person, e.g. 'I must have that doll', but used *c*.1600 with third person about someone else's compulsive desire; 2. are obliged to (*OED* 2). All editors have followed Rowe, emending to 'most'; it is true that an open 'o' is easily misread as 'u', but printers do not usually invent the less obvious reading and 'most' is relatively trite.
  haunt 1. resort to habitually (*OED* 3); 2. (of spirits) visit frequently. All members of the swallow family migrate annually and return to the same nesting-site; here the martlet becomes also the benign spirit of the castle.

10 **delicate** delightful, charming (with connotations of sensuous relaxation). This sense is still common in USA.

11–12 See Appendix A.

11–15 **the love … your trouble** Elaborately courtly phrasing, playing between the royal 'we' and the 'us' of common humanity. 'Our trouble' may mean either 'the trouble we cause' or 'loving attention can be troublesome': Duncan teaches her to thank him for being a nuisance as he loves her for the trouble she takes—and is troubled by it. See 1.4.45.

14 **God 'ield us** God pay, or reward (yield) us

17 **single** had also the sense of slight, poor, trivial (*OED* II.12b), derived from single as opposed to double cloth

21 **rest** remain
  hermits often used for beadsmen, whose business was to pray for the soul of another. Lady Macbeth may also mean that without Duncan's bounties, past and present, they would remain as obscure as hermits.

22 **coursed** chased (as hounds chase a hare)

23 **purveyor** one who makes preparation in advance, used especially of a domestic official who does this for a sovereign, which Duncan is inverting

24 **his great love … spur** i.e. urging him to speed, from the proverb 'he that hath love in his heart has spurs in his sides' (Dent L481)
  holp archaic past tense of 'help'

LADY MACBETH   Your servants ever
  Have theirs, themselves, and what is theirs, in count
  To make their audit at your highness' pleasure,
  Still to return your own.
DUNCAN   Give me your hand,
  Conduct me to mine host; we love him highly,   30
  And shall continue our graces towards him.
  By your leave hostess.   *Exeunt*

1.7   *Hautboys. Torches.*
     Enter a sewer and divers servants with dishes and
     service crossing over the stage.
     Then enter Macbeth

MACBETH
  If it were done when 'tis done, then 'twere well
  It were done quickly; if th'assassination
  Could trammel up the consequence and catch
  With his surcease, success, that but this blow
  Might be the be-all and the end-all—here,

27 count] F (compt)
   1.7.0.1 *Hautboys*] F (*Ho-boyes*)   0.3 crossing] CAPELL (*subst.*); *not in* F   5 end-all—here]
POPE; end all. Heere F; end-all here HANMER

27 **theirs** 1. their selves; 2. what is theirs
   in count on account
29 **Still to return your own** ready to return
   what is yours
32 **By your leave** a conventional apology for
   taking a liberty, here presumably kissing
   her
1.7.0.1–3 The initial dumb-show stresses
   the evening-time and the obligations of
   lavish hospitality. The general editor suggests that the hautboys might well have
   been played off-stage (as if in the banqueting hall), and that the torches might
   be placed by their bearers in cressets (iron
   holders) mounted on poles, or on the rear
   wall or pillars of the stage, leaving a clear
   stage for Macbeth's entry.
0.2 **sewer** official in charge of preparations
   and service of dinner
0.3 **service** all the materials needed to lay the
   table
1–28 The density of language is at its most
   complex here, ranging from the tongue-
   twisting ll. 1–7, enacting Macbeth's tortuous confusion, to the superb crescendo
   of ll. 16–25 playing out the angels'
   trumpets. At the same time a number of

metaphors, working at different levels of
imagination and vocabulary, are sustained simultaneously, often via a play
on words: the colloquial langauge of
fishing, horse-riding, and jumping; the
expository terminology of the law; and
the intensity of apocalyptic images. Commentary here is confined to individual
words and phrases: for fuller discussion,
see Introduction, pp. 7–10.
1–7 A single rhythmic unit though strictly
   in two sentences, the first ending 'end-all
   here', the second beginning 'here, | But
   here', so that the first 'here' belongs to
   both.
1–2 i.e. if the whole matter could be concluded with the single act of murder, it
   would be best to get it over with at once;
   see proverb 'the thing done is not to do',
   Dent T149.
3–5 Repeats and expands 1–2.
3 **trammel up** 1. catch (used of netting fish
   or birds); 2. fasten together the legs of a
   horse to prevent it straying
4 **surcease** making an end, i.e. murder
5 **end-all—here,** Punctuation cannot represent the double function of 'here' (see

But here, upon this bank and shoal of time,
We'd jump the life to come. But in these cases
We still have judgement here, that we but teach
Bloody instructions, which being taught, return
To plague th'inventor. This even-handed justice   10
Commends th'ingredience of our poisoned chalice
To our own lips. He's here in double trust:
First, as I am his kinsman, and his subject,
Strong both against the deed; then, as his host
Who should against his murderer shut the door,
Not bear the knife myself. Besides, this Duncan
Hath borne his faculties so meek, hath been
So clear in his great office, that his virtues
Will plead like angels, trumpet-tongued against
The deep damnation of his taking-off;   20
And pity, like a naked new-born babe,
Striding the blast, or Heaven's cherubim, horsed
Upon the sightless couriers of the air,
Shall blow the horrid deed in every eye

6 shoal] F (Schoole)   22 cherubim] F (Cherubin); Cherubins MUIR

note on ll. 1–7); F stresses the division of
sentences, but destroys the obvious
rhythm.

6 **bank and shoal** sand-bank and shallow
(F 'Schoole' was an alternative spelling,
surviving in 'school of porpoises'). Reference to the sea springs from the fishing
sense of 'trammel'.

7 **jump** leap over (from the horse sense of
'trammel'), hazard

8 **that** i.e. so that

10 **even-handed** impartial (emblems of Justice show a figure with equal weights in
each hand, or a balanced pair of scales)

11 **Commends** presents; there was also an
ecclesiastical sense, 'bestows'
ingredience 1. ingredients; 2. the fact or
process of entering in (*OED* 2). The
citations show a frequent (not invariable)
use of (2) in theological contexts, e.g. 'For
us in heaven to have ingredience' (*Sarum
Primer*, 1557); 'There is an ingrediency
(ingredience) and concurrence of all the
great and glorious perfection of God'
(John Weekes, *Truth's Conflict with Error*,
1650); this supports the allusion to the
communion cup in 'chalice'.
chalice 1. goblet; 2. cup used in the

communion service

17 **faculties** 1. personal powers (physical
and mental); 2. legal powers (of the king)

18 **clear** 1. pure; 2. shining (like light)

19 **plead** 1. beg; 2. advocate, as in law
courts

22 **Striding** 1. standing with legs apart; 2.
bestriding, as on a horse
blast 1. gust of air; 2. sound of trumpet;
the winds were sometimes pictured as
human figures blowing visible columns
of air
cherubim An order of angels, sometimes
represented as babies (like Cupids), sometimes as androgynous adolescents.
'Cherubim' is normally plural, earlier
spelt 'cherubin'; Muir's added 's' is unnecessary.

23 **sightless** 1. invisible; 2. blind (see 1.5.48)

24–5 **Shall . . . wind** The strongest sense is of
the paradox 'tears shall drown the
wind'—Macbeth's argument ends in impossibility, derived from proverbial 'small
rain allays great winds' (Dent R16).
Literal meanings have been suggested,
such as remission of breeze during a
shower, or eyes watering in a strong
wind, but they have scant relevance.

That tears shall drown the wind. I have no spur
To prick the sides of my intent, but only
Vaulting ambition, which o'erleaps itself
And falls on th' other—
    Enter Lady Macbeth
                How now? What news?

LADY MACBETH
He has almost supped: why have you left the
    chamber?

MACBETH
Hath he asked for me?

LADY MACBETH   Know you not he has?   30

MACBETH
We will proceed no further in this business.
He hath honoured me of late, and I have bought
Golden opinions from all sorts of people,
Which would be worn now in their newest gloss,
Not cast aside so soon.

LADY MACBETH   Was the hope drunk
Wherein you dressed yourself? Hath it slept since?
And wakes it now to look so green, and pale,
At what it did so freely? From this time,
Such I account thy love. Art thou afeard
To be the same in thine own act and valour   40
As thou art in desire? Wouldst thou have that
Which thou esteem'st the ornament of life,
And live a coward in thine own esteem,
Letting 'I dare not' wait upon 'I would',
Like the poor cat i'th' adage?

---

28 th' other—] ROWE; th'other. F, OXFORD   *Macbeth*] *not in* F

25–6 **I have ... intent** Reverting to the horse
    metaphor, in a simpler form.
26 **intent** 1. intention, purpose; 2. will
27 **Vaulting** 1. leaping (on to a horse's
    back); 2. leaping over a vaulting-horse
    (in a gym). Editors have offered a confla-
    tion 'vaulting into the saddle and falling
    over the other side', which seems to me
    absurd; the process is of the horse image
    dwindling from the splendour of the
    cherubim to a mere wooden horse as
    Macbeth's vision expires.
28 **th' other** 1. the other side (interrupted

by Lady Macbeth's entry); 2. otherness,
in the absolute sense of utter alienation.
F's full stop stresses (2) at the expense of
(1).
33 **sorts** kinds, ranks
37 **green, and pale** sick, i.e. with a hangover
41 **Wouldst thou have that** would you wish
    to have that
41–2 **that ... life** i.e. the crown
43 **And live** i.e. and yet live
44–5 The proverb 'the cat wanted to eat fish
    but dared not get her feet wet' (Dent
    C144).

MACBETH   Prithee peace:
  I dare do all that may become a man,
  Who dares do more is none.
LADY MACBETH   What beast was't then
  That made you break this enterprise to me?
  When you durst do it, then you were a man;
  And to be more than what you were, you would   50
  Be so much more the man. Nor time nor place
  Did then adhere, and yet you would make both—
  They have made themselves, and that their fitness now
  Does unmake you. I have given suck, and know
  How tender 'tis to love the babe that milks me;
  I would, while it was smiling in my face,
  Have plucked my nipple from his boneless gums
  And dashed the brains out, had I so sworn
  As you have done to this.
MACBETH   If we should fail?
LADY MACBETH   We fail?   60
  But screw your courage to the sticking place,
  And we'll not fail. When Duncan is asleep,
  Whereto the rather shall his day's hard journey

---

47 do] ROWE; no F   60 fail?] F; fail! ROWE

---

46 **become** 1. come to be, add up to; 2. be fitting to
    man As opposed to angel or beast (l. 47). The argument on 'man' constantly plays between 'decent humanity' and 'virile courage'.

47 **do** The emendation makes obvious sense; F's double negative does not.
    none i.e. more than *or* less than one, super- or sub-human

48 This line used to be quoted with absurd literalness as reason to suppose that an earlier scene had been cut, in which Macbeth proposed the murder. The same information provoked the same thought in both their minds, as they well knew it would.
    break reveal (as in 'break the news'), broach

52 **adhere** fit together (for murder)

53 **that their fitness** their very fitness

54–5 **I … me** Literal-mindedness has led to speculation on the number and paternity of Lady Macbeth's children: she is affirming her humanity in the same terms she used in 1.5.46–7.

57 **his** Commonly used for 'its'.

60–1 **We fail?** | **But** Seventeenth-century printers used '?' for both '?' and '!', which means there are two ways of reading, and playing, these words: 1. as an echo of Macbeth's question, probably sardonically incredulous, when 'but' = 'only'; 2. as a fatalistic acceptance— 'if we fail, we fail'. Actresses vary in their choice; Mrs Siddons first used (1) and later changed to (2).

61 **sticking place** the place where a screw cannot be turned further and sticks fast; the metaphor may derive from screwing back the cord on a crossbow.

63 **the rather** the sooner

Soundly invite him, his two chamberlains
Will I with wine and wassail so convince
That memory, the warder of the brain,
Shall be a fume, and the receipt of reason
A limbeck only; when in swinish sleep
Their drenchèd natures lies as in a death,
What cannot you and I perform upon   70
Th'unguarded Duncan? What not put upon
His spongy officers who shall bear the guilt
Of our great quell?

MACBETH   Bring forth men-children only:
For thy undaunted mettle should compose
Nothing but males. Will it not be received
When we have marked with blood those sleepy two
Of his own chamber, and used their very daggers,
That they have done't?

LADY MACBETH   Who dares receive it other,
As we shall make our griefs and clamour roar
Upon his death?

MACBETH   I am settled, and bend up   80
Each corporal agent to this terrible feat.
Away, and mock the time with fairest show,
False face must hide what the false heart doth know.

                                        Exeunt

69 lies] F1; lie F2

64 **chamberlains** chamber attendants on the
    king, here guards
65 **wassail** drinking healths
    convince overcome
66–8 **memory … limbeck only** The meta-
    phors are from alchemy, making a distil-
    lery of the drunken body: memory, the
    first ventricle of the brain, acted as war-
    der of the inner reason (for without
    memory there is no basis for thought); if
    that is reduced to vapour ('fume' was
    specifically used of noxious vapours sup-
    posed to rise from stomach to brain) then
    the receptacle ('receipt', which also im-
    plies orderly formulation) of reason (in an
    inner ventricle) will become only a lim-
    beck, which is the upper vessel of a still,
    into which vapourized spirits rise.

68 **swinish sleep** as in proverb 'as drunk as a
    swine' (Dent S1042)
69 **drenchèd** soaked
    lies singular for plural
72 **spongy** drunken
73 **quell** slaughter (the original meaning)
74–5 **For … males** The primary sense of
    masculinity is clear; there is a secondary,
    military, one, through puns on 'un-
    daunted' ('undented'), 'mettle' ('metal'),
    and 'males' ('mail' = armour).
75 **received** understood
79 **As** inasmuch as
80 **bend up** tauten the bow (as in l. 61)
81 **corporal agent** bodily instrument
82 **mock the time** befool the present world
    (see 1.5.62)

2.1 *Enter Banquo, and Fleance with a torch before him*

BANQUO How goes the night, boy?

FLEANCE The moon is down; I have not heard the clock.

BANQUO And she goes down at twelve.

FLEANCE I take't 'tis later, sir.

BANQUO

Hold, take my sword.—There's husbandry in Heaven,
Their candles are all out.—Take thee that too.—
A heavy summons lies like lead upon me,
And yet I would not sleep; merciful powers,
Restrain in me the cursèd thoughts that nature
Gives way to in repose.
    Enter Macbeth, and a Servant with a torch
            —Give me my sword.   10
Who's there?

MACBETH A friend.

BANQUO

What sir, not yet at rest? The King's abed.
He hath been in unusual pleasure,
And sent forth great largess to your offices.
This diamond he greets your wife withal

2.1.0 The direction is not clear: if Fleance carries the torch, it must be on a pole that he can put down when Banquo hands over his arms in ll. 5–6; alternatively 'torch' may mean a torchbearer, but if so there should be a comma after 'Fleance', and 'him' would probably refer to Banquo. I have retained F's punctuation, but the choices are a production matter. At l. 10.1 the torch is presumably carried by the servant, not by yet another supernumerary.

1–7 The opening lines, strictly prose, do suggest verse rhythm, though this is not fully established until l. 7.

5 **husbandry** domestic economy

6 **Their candles** proverbial phrase for the stars (the 'heavenly bodies'), Dent C49.1 **that** i.e. any other props or costume taken off to relax

7 **summons** i.e. urgent call to sleep

8–10 **merciful ... repose** The invocation is ambiguous in two ways: 1. Banquo may be literally appealing to beneficent spirits, or more vaguely using such langauge metaphorically to express his fears of nightmare, and so reflects Lady Macbeth's appeal to evil spirits in 1.5.39 ff.;

2. the 'cursed thoughts' may be his own ambitions, or his perception of Macbeth's. Muir capitalized 'powers' and quoted W. C. Curry, *Shakespeare's Philosophical Patterns* (Baton Rouge, La., 1937), 81, arguing a direct reference to the orders of angels, as in Milton, *Paradise Lost*, v. 601: 'Thrones, Dominations, Princedoms, Vertues, Powers'. This is possible, but unlikely because only the lower orders—archangels and angels—communicated directly with mankind from their adjacent positions in the chain of being. In any case *Shorter OED* (II.2) cites the exclamation 'merciful powers!' from 1596 with evidently similar vagueness to its modern use.

15 **largess to your offices** liberal gifts to the castle functionaries; 'office' was used for the office-holder (*OED* 4c)

16–20 The convoluted syntax of courtesy here and throughout the play tends to produce a merely technical obscurity that both contrasts with, and can be combined with, the complexity of fully imaginative language (see Introduction, p. 21).

16 **withal** 1. with; 2. in conclusion

By the name of most kind hostess, and shut up
In measureless content.
MACBETH   Being unprepared,
Our will became the servant to defect,
Which else should free have wrought.
BANQUO   All's well.   20
I dreamed last night of the three Weïrd Sisters;
To you they have showed some truth.
MACBETH   I think not of them:
Yet when we can entreat an hour to serve,
We would spend it in some words upon that business,
If you would grant the time.
BANQUO   At your kind'st leisure.
MACBETH
If you shall cleave to my consent, when 'tis,
It shall make honour for you.
BANQUO   So I lose none
In seeking to augment it, but still keep
My bosom franchised, and allegiance clear,
I shall be counselled.
MACBETH   Good repose the while.   30
BANQUO
Thanks sir, the like to you.   *Exit*
MACBETH
Go bid thy mistress, when my drink is ready,

2.1.21 Weïrd] F (weyward)   31 *Exit*] F (*Exit Banquo.*)

17 **shut up** The phrase is effectively in apposition to 'greets' despite the change of tense resulting from reported action. 'Shut up' was used for concluding affairs as well as for withdrawing to the confinement of a chamber.
18–20 i.e. 'at such short notice we could not achieve such perfect hospitality as we would have liked'
20 **wrought** Past participle of 'work', commonly in senses of creation or preparation (as in 'wrought-iron').
23–9 A series of allusions hint at Macbeth as king: Muir believed Macbeth was too good an actor to slip into the royal 'we', so glossed it in ll. 23–4 as 'you and I', and 'would' as 'should'; it is, of course, ambiguous, but the forms of royal utterance are unmistakable. They recur in l. 27, and in Banquo's response in ll. 27–8 until he finally articulates 'alle-

giance' in l. 29. Compare the insinuated conspiracy in 3.6.
25 **kind'st leisure** A possible phrase, though unusual; behind it lurks a pun with the obsolete senses of 'kind' (*OED, a.*2) 'belonging to one by birth; lawful, rightful' and (*OED, a.*3) 'rightful heir, etc.'; there seems also to be an aural play with 'at your highness' pleasure'.
26 **when 'tis** 1. when it is my leisure/pleasure; 2. when it is achieved, and I am king
29 **bosom franchised** 'bosom' was used of inner thoughts and feelings, 'franchised' for enfranchised, which strictly referred to freedom from servitude or bondage; Banquo clearly defends his integrity—the franchisement (like the allegiance after it) is simultaneously from his own corruption *and* from binding himself to a corrupt master, i.e. Macbeth.

She strike upon the bell. Get thee to bed.   *Exit Servant*
Is this a dagger which I see before me,
The handle toward my hand? Come, let me clutch
    thee.
I have thee not, and yet I see thee still.
Art thou not, fatal vision, sensible
To feeling as to sight? Or art thou but
A dagger of the mind, a false creation
Proceeding from the heat-oppressèd brain?   40
I see thee yet, in form as palpable
As this which now I draw.
Thou marshall'st me the way that I was going,
And such an instrument I was to use.
Mine eyes are made the fools o'th' other senses,
Or else worth all the rest. I see thee still;
And on thy blade and dudgeon gouts of blood,
Which was not so before. There's no such thing,
It is the bloody business which informs
Thus to mine eyes. Now o'er the one half world   50
Nature seems dead, and wicked dreams abuse
The curtained sleep; witchcraft celebrates
Pale Hecate's off'rings, and withered murder,

33 *Exit Servant*] F (*Exit.*)

34–50 For discussion of the dagger and its relation to other illusions in the play, see Introduction, p. 4.

37 **fatal** 1. decreed by fate, ominous; 2. deadly
sensible perceptible by the senses (here, of touch and sight)

40 **heat-oppressèd brain** Heat was thought of as a fluid substance which could literally weigh on the brain; it was also a property of the humours, producing passion and fever, thus figuratively oppressing the brain.

41 **palpable** tangible

42 Macbeth draws his actual dagger (hence, presumably, the short line leaving a pause for action); from this point on the speech refers to two daggers, one illusory, the other actual.

43 **marshal** usher, guide (*OED* 6)

45–6 **eyes ... rest** Since the senses are in conflict, either the other senses refute his sight, or sight alone is correct—an acute problem of the dependence of human knowledge on sensory perception.

47 **dudgeon** wooden handle
gouts large drops or splashes, especially used of blood (*OED*, *sb.*¹, II.5)

49 **informs** 1. gives form to; 2. teaches

50–1 **Now ... dead** Half the world is always in night, whose darkness seems like death.

52 **curtained** obscuring natural vision, literally by bed curtains, figuratively by eye-lids
sleep; witchcraft Several editors follow Davenant in inserting 'and' to cure a supposed deficiency in the line; Muir suggested that the pause was deliberate; I do not think there is any pause in speech other than that controlled by the rhythm and the semi-colon.
celebrates as in religious ritual (e.g. Mass)

53 **Pale Hecate** Hecate, goddess of witchcraft, was originally a moon-goddess (and so merged with Diana). Her name is disyllabic throughout the play (without final 'e' in F). There is little to associate the remote deity here with the goddess in 3.5 and 4.1.
off'rings ritual sacrifices

Alarumed by his sentinel, the wolf,
Whose howl's his watch, thus with his stealthy pace,
With Tarquin's ravishing strides, towards his design
Moves like a ghost. Thou sure and firm-set earth,
Hear not my steps, which way they walk, for fear
Thy very stones prate of my whereabout,
And take the present horror from the time   60
Which now suits with it. Whiles I threat, he lives;
Words to the heat of deeds too cold breath gives.
    A bell rings
I go, and it is done: the bell invites me.
Hear it not, Duncan, for it is a knell
That summons thee to Heaven, or to Hell.   *Exit*

2.2   *Enter Lady Macbeth*
LADY MACBETH
That which hath made them drunk, hath made me
    bold;

56 strides] POPE; sides F   57 sure] CAPELL (*conj.* Pope); sowre F   58 way they] ROWE; they
may F
  2.2.0 *Macbeth*] *not in* F

54 **Alarumed** called to arms, roused to
action
55 **Whose howl's his watch** Muir takes 'his'
to refer to wolf, and interprets the howl as
the wolf's watchword. This is possible,
though it seems easier to take 'his' as the
personified 'murder' alerted by the wolf's
howl. 'watch' can be the timepiece or the
watchman, originally a guard, later one
who patrolled the streets as guard and
also called the time.
56 **Tarquin's ravishing strides** Tarquin,
King of Rome, raped Lucrece, and in
Shakespeare's poem his walk from his
chamber to hers is amplified dramatically
in ll. 162–365, beginning with 'dead of
night' and 'wolves' death-boding cries'.
57 **sure** F's 'sowre' makes no obvious sense,
and the emendation is generally accepted.
57–61 **Thou sure ... with it** The ambiguities
of this passage rest largely with the
syntax: its simplest meaning depends on
understanding the subject of 'take' in
l. 60 as 'earth' which is the subject of
'hear not' in l. 58; thus if the earth hears
his steps, the stones will prate and their
echoes will unnerve him; his appeal in

this way is for reassuring silence from the
'sure and firm-set earth' so that he can
complete his walk to the murder while
the time suits it (it is, of course, now or
never). But if the subject of 'take' is
'stones' he seems rather to ask the stones
to take the horror away—i.e. prevent
him from committing the crime. At the
same time, as Warburton pointed out,
silence is the greatest horror, so that
noise would be reassuring—except that
this specific noise would still defeat his
intention. All ways round, the passage is
more eloquent of Macbeth's wish not to
murder Duncan than it is of his overt
appeal for help in doing so. The further
the lines are explored, the more endless
their contradictions—and that is precisely its achievement, as it was of
1.7.1–28; it will be its *effect*, however
they are delivered.

59 **stones prate** apparently proverbial, see
Dent S895.1, quoting George Gascoigne,
Posies (1575), i. 75: 'When men cry
mum and keep such silence long, | Then
stones must speak, else dead men shall
have wrong'

What hath quenched them, hath given me fire.—Hark,
      peace;
It was the owl that shrieked, the fatal bellman
Which gives the stern'st good night. He is about it,
The doors are open, and the surfeited grooms
Do mock their charge with snores. I have drugged their
      possets,
That death and nature do contend about them
Whether they live or die.
      Enter Macbeth
MACBETH Who's there? What ho?
LADY MACBETH

Alack, I am afraid they have awaked,   10
And 'tis not done; th'attempt, and not the deed,
Confounds us.—Hark—I laid their daggers ready,
He could not miss 'em. Had he not resembled
My father as he slept, I had done't.—My husband?
MACBETH
I have done the deed. Didst thou not hear a noise?
LADY MACBETH
I heard the owl scream, and the crickets cry.
Did not you speak?
MACBETH   When?
LADY MACBETH   Now.
MACBETH   As I descended?
LADY MACBETH   Ay.

8.1–9 *Enter Macbeth ... What ho?*] F; *Mac. [within] ... What ho!* STEEVENS, CAMBRIDGE; *Enter Macbeth above ... What ho? Exit* CHAMBERS   14 done't.—My husband?] F, *as two lines*; done't. *Enter Macbeth* My husband! CAMBRIDGE, CHAMBERS (*subst.*)

2.2.2 quenched 1. slaked their thirst; 2. extinguished (hence 'fire'); 3. stifled
3 **owl** a night-bird and commonly regarded as a harbinger of death (because of its weird cry); it was also a manifestation of Hecate, and of witches generally.
  fatal bellman The bellman was the nightwatchman; in Webster's *Duchess of Malfi* (1613–14) Bosola announces himself to the Duchess before he murders her as 'the common bellman | That usually is sent to condemn'd persons | The night before they suffer' (4.2.162–4) and presumably 'fatal bellman' is a reference to that practice, though I have not encountered other allusions to it.
4 **stern'st good night** i.e. the last, fatal one,

into death
5 **grooms** any male servants; more specifically used of royal attendants in the Lord Chamberlain's department
6 **possets** hot milk curdled with liquor, used as a cold cure and, presumably, a nightcap
9–14 The delay in Lady Macbeth's recognition of her husband has been much discussed; it may indicate that he speaks l. 9 offstage, or possibly 'above' (see l. 17) and enters the main stage in l. 14; but it may be part of enacting darkness on a daylight stage.
14 **My husband?** Possibly an exclamation; but if she mimes darkness, then it is uncertainty.

MACBETH
    Hark—who lies i'th' second chamber?
LADY MACBETH   Donalbain.
MACBETH
    This is a sorry sight.
LADY MACBETH   A foolish thought,
    To say a sorry sight.
MACBETH   There's one did laugh in's sleep,   20
    And one cried 'Murder', that they did wake each other.
    I stood, and heard them. But they did say their prayers,
    And addressed them again to sleep.
LADY MACBETH There are two lodged together.
MACBETH
    One cried 'God bless us', and 'Amen' the other,
    As they had seen me with these hangman's hands;
    List'ning their fear, I could not say 'Amen'
    When they did say 'God bless us'.
LADY MACBETH Consider it not so deeply.
MACBETH
    But wherefore could not I pronounce 'Amen'?   30
    I had most need of blessing, and 'Amen'
    Stuck in my throat.
LADY MACBETH   These deeds must not be thought
    After these ways: so, it will make us mad.
MACBETH
    Methought I heard a voice cry 'Sleep no more;
    Macbeth does murder sleep, the innocent sleep,
    Sleep that knits up the ravelled sleeve of care,

34–9 'Sleep … feast.'] HANMER; Sleep … Feast. F; 'Sleep … murder sleep' … feast. JOHNSON
36 sleeve] F; sleave MALONE

23 **addressed** in the original sense of 'prepared' (often specifically of clothing)
24 **two** It is not clear which two—it tends to
suggest the grooms, but that should not
be so; the usual suggestion is Donalbain
and someone else (not Malcolm, or she
should have mentioned him in l. 18). The
reference may be careless, and in any
case is not important. (See also Appendix
A, ll. 19–24.)
26 **As** as if
hangman used of the executioner, whatever means he employed; part of his
duties was to disembowel and quarter the
hanged man.

34–9 F uses no inverted commas, and editors vary in their placing; the whole
passage is a formal apostrophe to Sleep,
in the mode of Sidney's and Daniel's well-
known sonnets (*Astrophel and Stella* 39;
Delia 54) and so is only vaguely associated with the illusion of an actual voice;
it cannot be divided sensibly in delivery.
36 **ravelled sleeve** 1. frayed out sleeve (as in
care-worn, out-at-elbow knitting); 2.
'sleave' is a thin filament of silk made by
ravelling a thicker thread; most editors
give this as the primary meaning here
although they stress the clothing imagery in the play—'sleave' seems to me

The death of each day's life, sore labour's bath,
Balm of hurt minds, great nature's second course,
Chief nourisher in life's feast.'

LADY MACBETH　　　　　　　　　　What do you mean?

MACBETH

Still it cried 'Sleep no more' to all the house:　　　　　40
'Glamis hath murdered sleep, and therefore Cawdor
Shall sleep no more—Macbeth shall sleep no more.'

LADY MACBETH

Who was it that thus cried? Why worthy thane,
You do unbend your noble strength to think
So brain-sickly of things—go get some water,
And wash this filthy witness from your hand.
Why did you bring these daggers from the place?
They must lie there—go carry them, and smear
The sleepy grooms with blood.

MACBETH　　　　　　　　　　　　I'll go no more:
I am afraid to think what I have done;　　　　　　　50
Look on't again, I dare not.

LADY MACBETH　　　　　　　　　　Infirm of purpose;
Give me the daggers; the sleeping, and the dead,
Are but as pictures; 'tis the eye of childhood
That fears a painted devil. If he do bleed,
I'll gild the faces of the grooms withal,
For it must seem their guilt.　　　　　　　　　*Exit*

　　　　Knock within

MACBETH　　　　　　　　　　Whence is that knocking?
How is't with me, when every noise appals me?
What hands are here? Ha, they pluck out mine eyes.

40–2] HANMER; *without inverted commas* F

of doubtful relevance. A further sense of
'ravelled' (noun) is of wholemeal bread;
irrelevant in this line, it may yet have
suggested 'nourisher' in l. 39 (not in *OED*,
but Muir cites Harrison, 'England', in Holinshed (ed. Furnivall, 1877, i. 154), 'The
raueled is a kind of cheat bread').

37 **bath** The 1591 edition of *Astrophel and
Stella* printed l. 2 of Sonnet 39 as 'The
bathing place of wit ...', where later
editions have 'baiting'; Sidney may, or
may not, have been punning, but this is

surely the origin of Shakespeare's unex-
pected image.

38 **second course** 1. second main way (the
other being wakeful activity); 2. second
course of a meal (hence l. 39)

44 **unbend** relax (originally used of a bow)

53–4 **'tis the eye ... painted devil** Dent B703
refers this to the saying 'bugbears to scare
children'.

55–6 **gild ... guilt** Old gold was red, hence
'gilt' is blood on the painted devils she
makes of the grooms.

57 **appals** 1. makes pale; 2. shocks

Will all great Neptune's ocean wash this blood
Clean from my hand? No—this my hand will rather   60
The multitudinous seas incarnadine,
Making the green one red.

    Enter Lady Macbeth

LADY MACBETH

My hands are of your colour, but I shame
To wear a heart so white.

    Knock

                   I hear a knocking
At the south entry: retire we to our chamber;
A little water clears us of this deed.
How easy is it then! Your constancy
Hath left you unattended.

    Knock

                 Hark, more knocking.
Get on your nightgown, lest occasion call us
And show us to be watchers; be not lost   70
So poorly in your thoughts.

---

62 green one red] F4; Greene one, Red F1; green, one red JOHNSON   62.1 *Macbeth*] *not in* F

59–60 **Will ... hand** Tilley (W85) gave 'all the water in the sea cannot wash out this stain' as a proverb, but Dent suggests it may not have been proverbial and cites instead *Saint Bernard His Meditations* (3rd edn. 1614), Med. XII (addressed to Pilate): 'Well might a little water clear the *spots* of thy *hands*, but all the water in the *Ocean* could not wash away the blots of thy soul.' It is hard to believe the idea was not a commonplace although all later uses seem to derive from *Macbeth*.

61–2 The astonishing shift of language from the polysyllabic Latinate l. 61 to the monosyllabic English l. 62 is in literal sense virtually repetition: hence 'green one' corresponds to 'multitudinous seas'. F1 read 'green one, red' intelligibly enough; F4 dropped the comma, and Johnson, objecting to 'green one', re-punctuated 'green, one red'. Many editors follow Johnson, glossing 'one red' as 'all red', although it obviously violates the rhythm of the cadence; actors conscientiously attempting it seem always to need a strained pause after 'green'. 'Green one' has the obvious second sense 'innocent'.

64 **white** i.e. cowardly

67–8 **Your constancy ... unattended** An ironic phrase which is slightly obscure; it seems to be a conflation of two constructions: 1. 'Your constancy [i.e. firmness of purpose] has deserted you'; 2. 'that has left you unattended [i.e. unguarded]'. Thus, as Foakes says, the general sense is 'you have lost your nerve' with her usual implication of 'pull yourself together', backing up her assertion of cool practicality.

69 **nightgown** dressing-gown (bed garments were not worn, and the word was not applied to them before the nineteenth century)

    occasion call us the circumstances (discovery of the murder) cause us to be summoned (while we are still fully dressed)

70 **watchers** ones who keep watch, i.e. are about while others sleep

MACBETH

To know my deed, 'twere best not know myself.

Knock

Wake Duncan with thy knocking: I would thou
couldst.   *Exeunt*

2.3   *Enter a Porter. Knocking within*

PORTER Here's a knocking indeed: if a man were porter of
Hell gate, he should have old turning the key.

Knock

Knock, knock, knock. Who's there i'th' name of Beelzebub? Here's a farmer, that hanged himself on th'
expectation of plenty; come in time—have napkins
enough about you, here you'll sweat for't.

Knock

Knock, knock. Who's there in th' other devil's name?
Faith, here's an equivocator, that could swear in both
the scales against either scale, who committed treason

72.1 *Knock*] F (*after* deed)
   2.3.5 time] F; farmer CAMBRIDGE; time-server WILSON   6 enough] F (enow)

---

72 i.e. 'if I must acknowledge my crime I
must forget the man I have been, and
am'—he is condemned to self-division
2.3.0 *within* off-stage (inside the tiring-
house behind the playing area)
1–2 **porter of Hell gate** In some miracle
plays a comic demon, matching St Peter
on the gate of Heaven: this porter is
setting up a well-known comic scene, see
Introduction, pp. 79–81.
2 **have old turning the key** have too much
turning of the key to do (letting in so
many evil spirits). 'old' was used for
'great' or 'too much', as now in phrases
like 'high old time'.
3–4 **Beelzebub** A popular devil name, one
of the very few found in the Bible; see
Introduction, pp. 80–1.
4–5 **farmer ... plenty** Apparently not
proverbial, but a popular joke against
early capitalism: the farmer hoards grain
when it is short and badly needed, hoping
for a further price rise; but if the next
harvest is plentiful, the price falls and he
is ruined—hence his suicide. Malone
quoted Joseph Hall, *Satires* (1597–8), iv.
6, 'Ech Muck-worme will be rich with

lawlesse gaine, | Altho' he smother up
mowes of seven yeares graine, | And
hang'd himself when corne grows cheap
again.' The protesting populace riots
about grain-hoarding in Act 1 of *Coriolanus.*
5 **come in time** i.e. a timely entry (the fire's
all ready). Emendation is common but
unnecessary.
napkins needed to mop up the sweat;
associated with the delicate feeding
habits of the rich.
7 **th' other devil's name** The Porter does
not know the name because in miracle
plays attendant devils were usually
anonymous. See Introduction, pp. 80–1.
8 **equivocator** The Jesuit Father Garnet
justified his use of equivocation under
interrogation, at his trial for the gunpowder plot in March 1606. See Introduction, pp. 59–60. Garnet used the name of
'Farmer' which may have triggered this
reference, but it is anyway pertinent to
the constantly equivocal language of the
play.
8–9 **both ... scale** i.e. pervert the balance of
justice's equal scales.

enough for God's sake, yet could not equivocate to   10
Heaven: O come in, equivocator.
    Knock
Knock, knock, knock. Who's there? Faith, here's an
English tailor come hither, for stealing out of a French
hose: come in, tailor, here you may roast your goose.
    Knock
Knock, knock. Never at quiet—what are you?—But this
place is too cold for Hell. I'll devil-porter it no further: I
had thought to have let in some of all professions, that
go the primrose way to th' everlasting bonfire.
    Knock
Anon, anon, I pray you remember the porter.
                           ⌈*Opens door*⌉
    Enter Macduff, and Lennox
MACDUFF
Was it so late, friend, ere you went to bed,   20
That you do lie so late?
PORTER Faith sir, we were carousing till the second cock;
and drink, sir, is a great provoker of three things.
MACDUFF What three things does drink especially pro-
voke?
PORTER Marry sir, nose-painting, sleep, and urine. Lech-
ery, sir, it provokes, and unprovokes: it provokes the

19 *Opens door*] not in F

10 **for God's sake** The persecuted Catholics
justified equivocation on religious
grounds; the Protestant authorities saw
it as damnable perjury.
13–14 **English tailor ... French hose** An old
joke against tailors, skimping on the
cloth where they could get away with it,
which would be easy with French hose
which was either very baggy and long, or
short and tight (Philip Stubbes, *Anatomie
of Abuses* (1585), p. 23b, describes both).
There may also be an allusion to the
'French disease', gonorrhea or syphilis,
especially uncomfortable in tight
breeches (see 'goose' below).
14 **roast your goose** i.e. cook your goose, do
for you; 'goose' was also a term for
venereal disease, and for a tailor's
smoothing iron.
16 **devil-porter it** i.e. play out the role
17 **of all professions** Suggesting a play at-

tacking all trades, a common satiric idea.
18 **primrose way** 'primrose' was often used
for 'finest' or 'best', but the association
with a path of gaudy pleasure seems to be
peculiar to Shakespeare; see *Hamlet*
1.3.49–50, 'a puffed and reckless liber-
tine | Himself the primrose path of dalli-
ance treads', and *All's Well* 4.5.53–5
('flow'ry way')
19 **remember the porter** i.e. give him a tip as
he opens the door; this is commonly said
to be addressed to the audience, for
whom it is the moral on the end of his
little play; but it is obviously aimed at
Macduff.
22 **second cock** the full dawn (first cockcrow
is early dawn); in *Romeo* 4.4.3–4, it is
three o'clock
25–34 The bawdy puns on drink and lechery are familiar still, and self-explanatory
for the most part.

desire, but it takes away the performance. Therefore
much drink may be said to be an equivocator with
lechery: it makes him, and it mars him; it sets him on,   30
and it takes him off; it persuades him, and disheartens
him; makes him stand to, and not stand to—in conclusion, equivocates him in a sleep, and giving him the lie,
leaves him.

MACDUFF I believe, drink gave thee the lie last night.

PORTER That it did, sir, i'the very throat on me; but I
requited him for his lie and, I think, being too strong for
him, though he took up my legs sometime, yet I made a
shift to cast him.

          Enter Macbeth

MACDUFF

    Is thy master stirring?   40
    Our knocking has awaked him: here he comes.

LENNOX (*to Macbeth*)

    Good morrow, noble sir.

MACBETH   Good morrow both.

MACDUFF

    Is the King stirring, worthy thane?

MACBETH   Not yet.

39.1] F; *after* 38 COLLIER

30 **it makes … mars him** proverbial,
    Dent M48
32 **stand to … stand to** Various military and
    bawdy senses contribute to this: 1. pre-
    sent a bold front to (an enemy) (*OED* e);
    2. toil without flagging at painful or
    severe labour (*OED* c); 3. stand with
    one's weapon in readiness for action
    (*OED* d)—in a sexual sense this is still
    used about breeding animals.
33 **equivocates him in a sleep** 1. cheats him
    into sleep; 2. deceives him into a dream
    giving him the lie 1. calling him a liar; 2.
    forcing him to lie down; 3. making him
    urinate ('lye', strictly a detergent, was
    used for urine)
36 **i'the very throat on me** 1. lie in my throat
    (Dent T268), meaning completely de-
    ceive—possibly because if you killed
    somebody while they were lying they
    went to Hell; see Hamlet's refusal to kill

Claudius in 3.3.84–96 because he is
praying and would go to Heaven, and
Nashe's grotesque elaboration of the idea
at the end of *The Unfortunate Traveller*
(1594); 2. the drink lies in his throat till
he vomits.
38 **took up my legs** 1. as a wrestler effects a
    fall; 2. made me legless
38–9 **made a shift to cast him** 1. was content
    to vomit; 2. managed to throw off the
    effects of drink (as a wrestler escapes from
    a hold)
39.1 The common emendation looks plaus-
    ible for proscenium arch theatres, but
    actors entering from upstage are not
    visible to those on stage until they come
    down or call attention to themselves.
    There is no exit for the Porter, who may
    linger, slink away, or be dismissed by a
    gesture from Macbeth.

MACDUFF

He did command me to call timely on him,
I have almost slipped the hour.

MACBETH I'll bring you to him.

MACDUFF I know this is a joyful trouble to you: but yet 'tis
one.

MACBETH The labour we delight in physics pain—this is
the door.   50

MACDUFF I'll make so bold to call, for 'tis my limited
service.   *Exit Macduff*

LENNOX Goes the King hence today?

MACBETH He does: he did appoint so.

LENNOX

The night has been unruly: where we lay
Our chimneys were blown down, and, as they say,
Lamentings heard i'th' air; strange screams of death,
And prophesying with accents terrible
Of dire combustion and confused events
New-hatched to th' woeful time. The obscure bird   60
Clamoured the livelong night. Some say the earth
Was feverous and did shake.

MACBETH   'Twas a rough night.

LENNOX

My young remembrance cannot parallel
A fellow to it.
        Enter Macduff

60 time.] F; time, HANMER

44 **timely** early
45–52 See Appendix A.
45 **slipped the hour** let the time slip by, failed
in keeping the hour (*OED* IV.20b and c)
49 **The labour ... pain** From the sententious
'what we do willingly is easy' Dent D407.
physics purges, treats with medicine
51–2 **limited service** appointed duty
59 **combustion** fire (possibly from lightning).
Foakes, following Muir, glosses 'tumult',
referring to *Henry VIII* 5.4.48; but in his
Arden edition of *Henry VIII* he glosses
that 'commotion' referring to this line in
Macbeth; the phrase there is 'kindling
combustion' which must refer to fire—if
there is a figurative sense in either context it should surely be 'explosion'.
60 **New-hatched to th' woeful time** This

phrase has caused mystification; it seems
to me clearly to be a prophecy of the
violent changes future time would give
birth to.
obscure bird bird of darkness, the ominous owl. By placing a comma after 'time',
Hudson (conj. Johnson) made the bird
utter the prophecy; the owl's clamour is
a repeated inarticulate cry of doom, as
in 2.2.3.
62 **feverous** feverish
63 **remembrance** memory, with the specific
sense of the period over which one's
memory extends (*OED* 3b)
64–6 The lines have often been regularized
(see Appendix A); but Macduff's violent
entry demands a break in the rhythm.

MACDUFF

O horror, horror, horror!

Tongue nor heart cannot conceive, nor name thee.

MACBETH *and* **LENNOX** What's the matter?

MACDUFF

Confusion now hath made his masterpiece:

Most sacrilegious murder hath broke ope

The Lord's anointed temple, and stole thence   70

The life o'th' building.

MACBETH

What is't you say, the life?

LENNOX   Mean you his majesty?

MACDUFF

Approach the chamber, and destroy your sight

With a new Gorgon. Do not bid me speak—

See, and then speak yourselves—

                       Exeunt Macbeth and Lennox

                       Awake, awake,

Ring the alarum bell; murder and treason,

Banquo and Donalbain—Malcolm awake,

Shake off this downy sleep, death's counterfeit,

And look on death itself: up, up, and see

The great doom's image—Malcolm, Banquo,   80

---

75 *Exeunt ... Lennox*] F (*at end of line*)

66 **conceive** 1. take into the mind (from Latin *concipere,* meaning to take in and hold); 2. create, as in the womb
name The act of naming conferred actuality on a thing otherwise amorphous; in Genesis 2 Adam named the individual species of plants and animals, which were previously generic only. In one theory of language, it had this magical power, to which St John referred in the beginning of his Gospel, 'In the beginning was the Word'; thus the conception of the deed becomes actual only when it is named.

69–71 **sacrilegious ... temple ... building** The king's 'two bodies' were first his human one, and secondly his divine one, instituted by his anointing as king, God's representative on earth, whose life was therefore sacred. See *Richard II* 4.1.238–240: 'I find myself a traitor with the rest, | For I have given here my soul's consent | T'undeck the pompous body of a king'.

72 Muir suggests that Macbeth and Lennox speak simultaneously, which is possible; but their words do follow sequentially, and there may be a pause after Macduff's speech.

73–4 **destroy ... Gorgon** Medusa was the mortal one of the three Gorgon sisters whose heads grew snakes and whose teeth were tusks; anyone who looked at them was turned to stone. Perseus averted his eyes and cut her head off, thereafter using it effectively against his enemies.

78 **death's counterfeit** sleep as an image of the death it resembles was common, Dent S527

80 **The great doom's image** Doomsday, the day of judgement, is the ultimate death of all, when the dead rise from their graves (l. 76) and are eternally assigned

As from your graves rise up, and walk like sprites,
To countenance this horror. Ring the bell.
> *Bell rings.*
> *Enter Lady Macbeth*

LADY MACBETH
What's the business, that such a hideous trumpet
Calls to parley the sleepers of the house?
Speak, speak.

MACDUFF   O gentle lady,
'Tis not for you to hear what I can speak:
The repetition in a woman's ear
Would murder as it fell.
> *Enter Banquo*

                O Banquo, Banquo,
Our royal master's murdered.

LADY MACBETH   Woe, alas:
What, in our house?

BANQUO   Too cruel, anywhere.   90
Dear Duff, I prithee contradict thyself,
And say it is not so.
> *Enter Macbeth and Lennox*

---

82 Ring the bell] F; *as s.d.* THEOBALD   82.2 *Macbeth*] *not in* F   92.1] CAPELL; *Enter Macbeth,*
Lenox, and Rosse F

to Heaven or to Hell. It is prefigured by
individual death as that is prefigured by
sleep.

81 **sprites** spirits, i.e. ghosts
82 **countenance** keep in countenance (*OED*
    6), i.e. be in keeping with; editors sug-
    gest, reasonably, that it also means 'be-
    hold' (and in that sense 'know')
    Ring the bell It has often been suggested
    that this was a book-keeper's note, crept
    into the text; but since Macduff's order in
    l. 76 has not yet been obeyed, his im-
    patience is intelligible.
83–5 See Appendix A.
83–4 **trumpet ... parley** The trumpet which
    seems to extend the references to the last
    judgement can also be only a summons
    to meet.
87–90 **in a woman's ear ... What, in our**
    house? ... Too cruel anywhere Macduff's
    conventional assumptions about gender

roles transpose into hers as hostess,
which earn Banquo's rebuke. Comments
on this exchange often ignore her actual
role and perfect knowledge of events.
92.1 Ross should not enter with Macbeth
    and Lennox since he did not go with them
    to view the body, and was not on stage to
    do so; he cannot be on stage to the end of
    the scene because he must make a fresh
    entry for 2.4. Before this scene he had
    always been twinned with Lennox, and
    F's direction may be an accident due to
    copyist or compositor. Oxford suggests
    that he may be required to help Lady
    Macbeth off at l. 127, which would still
    enable him to enter for 2.4; if so, his
    entry here may be an afterthought (he
    does not speak at all), noted marginally
    and wrongly printed here. But he is not
    actually needed in the scene at all (see
    note to l. 127).

MACBETH

　Had I but died an hour before this chance,
　I had lived a blessèd time; for from this instant
　There's nothing serious in mortality—
　All is but toys: renown and grace is dead,
　The wine of life is drawn, and the mere lees
　Is left this vault to brag of.

　　　　Enter Malcolm and Donalbain

DONALBAIN

　What is amiss?

MACBETH　　　　You are, and do not know't:
　The spring, the head, the fountain of your blood　　100
　Is stopped, the very source of it is stopped.

MACDUFF

　Your royal father's murdered.

MALCOLM　　　　　　O, by whom?

LENNOX

　Those of his chamber, as it seemed, had done't:
　Their hands and faces were all badged with blood,
　So were their daggers which, unwiped, we found
　Upon their pillows; they stared, and were distracted,
　No man's life was to be trusted with them.

MACBETH

　O yet I do repent me of my fury,
　That I did kill them.

MACDUFF　　　　Wherefore did you so?

MACBETH

　Who can be wise, amazed, temp'rate, and furious,　　110
　Loyal and neutral, in a moment? No man:
　Th'expedition of my violent love
　Outran the pauser, reason. Here lay Duncan,

---

113 Outran] JOHNSON; Out-run F

93–8 Macbeth's speech is so much in the language of his soliloquies that its use here as public address creates a curious ambiguity between what it means to his stage audience and what it means to the theatre audience.
97 **lees** sediment, dregs
98 **vault** 1. wine vault; 2. sky (possibly 3. burial vault)
100–2 Macbeth's resumption of courtly convolution is bluntly displaced by Macduff's plain language.

104 **badged** marked as with an emblem, i.e. of guilt
112 **expedition** the action of expediting, prompt or hasty action
113 **Outran** It is hard to see why most editors have ignored Johnson's emendation; Oxford comments 'OED does not record "run" as a past tense; misreading would be easy'.
　　pauser one who pauses to think (also with implications of cowardice)

His silver skin laced with his golden blood,
And his gashed stabs looked like a breach in nature
For ruin's wasteful entrance; there the murderers,
Steeped in the colours of their trade, their daggers
Unmannerly breeched with gore—who could refrain,
That had a heart to love, and in that heart,
Courage, to make's love known?

LADY MACBETH   Help me hence, ho!   120

MACDUFF   Look to the lady.

MALCOLM (*aside to Donalbain*)
Why do we hold our tongues, that most may claim
This argument for ours?

DONALBAIN (*aside to Malcolm*)
                      What should be spoken here,
Where our fate, hid in an auger-hole, may rush
And seize us? Let's away, our tears are not yet brewed.

MALCOLM (*aside to Donalbain*)
Nor our strong sorrow upon the foot of motion.

BANQUO   Look to the lady—   *Exit Lady Macbeth, helped*
And when we have our naked frailties hid,
That suffer in exposure, let us meet

---

114 **laced** patterned with streaks of colour

115–16 **a breach ... entrance** The metaphor is from breaching the walls of a besieged city, through which the invading army causes ruin.

116 **wasteful** that causes devastation

116–18 **there ... gore** The general sense is clear, but the specific metaphoric allusions are elusive. 'Steeped' had a sense of soaking in blood (*OED* 2b); 'breeched' means clothed in breeches, but was also used of flogging (drawing blood), so that the image is 'unmannerly' both because the breeches were filthy and because they were pulled down to expose naked buttocks; 'the colours of their trade' are the specific badges or uniforms of a given trade, in this case murderers, and the image carries also the (false) implication that they were professionals and, as tradesmen, social inferiors. Macbeth thus reduces them first to the status of those he hires in 3.1 (see especially l. 92), and then to flogged criminals, or possibly small boys.

120 **Help me hence, ho!** It is ambiguous whether Lady Macbeth is pretending, or does actually faint—and will inevitably be so in performance, despite the many editors who have pronounced one way or the other.

122–6 Malcolm and Donalbain evidently speak aside while business takes place around Lady Macbeth; Banquo's repetition of Macduff's order serves presumably as recall to the audience whose attention has been distracted: the effect is of simultaneous action. See Appendix A (121–6).

124 **auger-hole** auger is a carpenter's tool for boring small holes in wood—hence a virtually invisible space (possibly alluding to a dagger-stab)

125 **brewed** fully fermented, ready for pouring

126 **upon the foot of motion** up on its feet for movement, ready to move

127 See note on ll. 122–6. Lady Macbeth must either be helped by servants answering Macduff's call, or by Lennox or, if he is available (see note to 92.1), by Ross.

128 **our naked frailties hid** 1. dressed fully; 2. got over our immediate distress

And question this most bloody piece of work,   130
To know it further. Fears and scruples shake us:
In the great hand of God I stand, and thence
Against the undivulged pretence I fight
Of treasonous malice.

MACDUFF   And so do I.

ALL   So all.

MACBETH
Let's briefly put on manly readiness,
And meet i'th' hall together.

ALL   Well contented.

Exeunt all but Malcolm and Donalbain

MALCOLM
What will you do? Let's not consort with them—
To show an unfelt sorrow is an office
Which the false man does easy. I'll to England.

DONALBAIN
To Ireland, I. Our separated fortune   140
Shall keep us both the safer; where we are,
There's daggers in men's smiles—the near in blood,
The nearer bloody.

136.1 all ... Donalbain] *not in* F   142 near] F (neere); near' DELIUS; nea'er HUNTER, OXFORD

131 **scruples** uncertainties, of the truth and
of how to act
132–4 Banquo reverts to the convoluted
obscurity of courtly language, now
loaded with the impossibility of candour.
132 **In the great hand of God I stand** I cannot
find analogues for 'the great hand of
God', though it obviously alludes to his
all-knowing, all-protecting reassurance.
Isaiah 49: 16–17 may be relevant: 'Behold, I have graven thee upon the palm of
mine hands: thy walls are ever in my
sight. | Thy builders make haste: thy
destroyers and they that made thee waste
are departed from thee'. An annotated
edition of the Bishops' Bible (1599)
glosses 'mine hands' 'Because I would
not forget thee'; 'thy walls' 'Meaning,
the good order of policy and discipline';
'make haste' 'I have a continual care to
build thee up again, and to destroy thine
enemies'—all of which fit this context. By
contrast 'I stand', which is a cliché for the
moral soldier, becomes an oddly grotesque image (like Gulliver in Brobding-
nag) of Banquo's bewildered impotence
and futile defiance. None the less the

image impacts rather as an accident of
mixed clichés than as an imaginative
perception.
133 **undivulged pretence** hidden purpose (of
further crime); but as no one has yet
claimed the crown, there is a hint of
'pretence' as 'claim' to the throne
134 ALL i.e. Macbeth, Malcolm, and Donalbain; and Lennox if he is still on stage (see
note on l. 127)
135 **manly readiness** full preparation, i.e.
dressed in clothes and weapons
136 ALL Macduff, Banquo, Malcolm, Donalbain (and possibly Lennox as at l. 134)
137–43 See Appendix A.
137 **consort** associate
142 **near** F's 'neere' may derive from a
contraction of 'nearer', but emendation
hardly helps what is anyhow a clumsy
epigram.
142–3 **the near ... bloody** those closest in
blood (to Duncan) are most in danger.
Tilley thought this related to the proverb
'the nearer in kin, the less in kindness'
(K38; *Hamlet* 1.2.65); Dent disagreed,
but ironic inversion may explain the
obscurity.

MALCOLM   This murderous shaft that's shot
Hath not yet lighted, and our safest way
Is to avoid the aim. Therefore to horse,
And let us not be dainty of leave-taking,
But shift away: there's warrant in that theft
Which steals itself when there's no mercy left.

Exeunt

2.4.   *Enter Ross, with an Old Man*

OLD MAN
Threescore and ten I can remember well,
Within the volume of which time I have seen
Hours dreadful and things strange; but this sore night
Hath trifled former knowings.

ROSS   Ha, good father,
Thou seest the heavens, as troubled with man's act,
Threatens his bloody stage. By th' clock 'tis day,
And yet dark night strangles the travelling lamp;
Is't night's predominance, or the day's shame,
That darkness does the face of earth entomb
When living light should kiss it?

OLD MAN   'Tis unnatural,   10
Even like the deed that's done. On Tuesday last,
A falcon tow'ring in her pride of place
Was by a mousing owl hawked at, and killed.

143–5 **This murderous ... aim** i.e. the mur-
    derer's plan is not yet complete and we'd
    best avoid the obvious danger
144 **lighted** alighted, landed
146 **dainty** particular
147 **shift away** get away, with a sense of
    secrecy or evasion
    warrant justification
148 **steals** Punning on theft and 'stealing
    away', so that the theft steals away from
    itself.
2.4.3 **sore** severe, violent
    4 **Hath ... knowings** made previous experi-
    ence seem trivial
    father a respectful title for an old man
5–6 **heavens ... act ... stage** 'heavens' is
    the sky and, when personified, the gods;
    it is also the canopy over the stage at the
    Globe Theatre (symbolically adorned
    with sky and stars), and thus relates to
    'act' and 'stage'.

6–10 **By th' clock ... kiss it** The whole
    metaphor depends on personifying Night
    and Day.
7 **strangles** in the wider sense of 'smothers'
    travelling lamp sun, 'travelling' as it
    crosses the sky, or as it lights human
    travellers; 'travelling' can also be 'trav-
    ailing', i.e. suffering, or 'labouring' to
    become visible: the opposition of death
    and life (as well as shame) in ll. 7–10
    suggests the use of 'travail' for childbirth.
8 **predominance** in the astrological sense of
    superior influence
    day's shame i.e. at Duncan's murder
12 **tow'ring ... pride of place** Terms in
    falconry: 'towering' is the falcon's up-
    ward spiralling flight; 'pride of place' its
    highest point where it hovers before
    diving on its prey.
13 **mousing owl** Unlike the falcon, the owl
    flies low when hunting.

ROSS

And Duncan's horses—a thing most strange, and
    certain—
Beauteous and swift, the minions of their race,
Turned wild in nature, broke their stalls, flung out,
Contending 'gainst obedience, as they would
Make war with mankind.

OLD MAN   'Tis said they ate each other.

ROSS

They did so, to th' amazement of mine eyes
That looked upon't.
        Enter Macduff
                          Here comes the good Macduff.   20
How goes the world, sir, now?

MACDUFF   Why, see you not?

ROSS

Is't known who did this more than bloody deed?

MACDUFF

Those that Macbeth hath slain.

ROSS   Alas the day,
What good could they pretend?

MACDUFF   They were suborned;
Malcolm and Donalbain, the King's two sons,
Are stol'n away and fled, which puts upon them
Suspicion of the deed.

ROSS   'Gainst nature still—
Thriftless ambition, that will raven up
Thine own life's means. Then 'tis most like
The sovereignty will fall upon Macbeth.   30

MACDUFF

He is already named, and gone to Scone

2.4.18 ate] F (eate)

15 **minions** darlings, favourites—i.e. best
   cared-for and finest
24 **pretend** aspire to
   suborned procured (by bribery, etc.)
28 **Thriftless** unprofitable, wasteful
   raven up devour voraciously
29 **Thine own life's means** what sustains
   you: their father as provider; but here
   also the wastefulness is that a murder
   supposed to gain the crown has forced
   their flight

30 The Scottish monarchy was elective (by
   the thanes), though (as Malcolm's position makes clear) primogeniture was
   usually respected. Macbeth's claim could
   be as triumphant general, but also that,
   once Malcolm and Donalbain were disqualified, he was Duncan's nearest surviving relative (first cousin).
31 **Scone** An abbey just north of Perth
   which housed the sacred stone on which
   the Scottish kings were crowned.

To be invested.

ROSS

Where is Duncan's body?

MACDUFF   Carried to Colmekill,

The sacred storehouse of his predecessors

And guardian of their bones.

ROSS   Will you to Scone?

MACDUFF

No cousin, I'll to Fife.

ROSS   Well, I will thither.

MACDUFF

Well, may you see things well done there—adieu,

Lest our old robes sit easier than our new.

ROSS Farewell, father.

OLD MAN

God's benison go with you, and with those   40

That would make good of bad, and friends of foes.

                                        Exeunt

3.1   *Enter Banquo*

BANQUO

Thou hast it now, King, Cawdor, Glamis, all,

As the Weïrd Women promised, and I fear

Thou played'st most foully for't; yet it was said

It should not stand in thy posterity,

But that myself should be the root and father

Of many kings. If there come truth from them,

As upon thee, Macbeth, their speeches shine,

Why by the verities on thee made good

May they not be my oracles as well,

---

37 Well,] THEOBALD; Well F   41 *Exeunt*] F (*Exeunt omnes*)
3.1.2 Weïrd] F (weyard)

---

32 **invested** clothed in the insignia of monarchy
33 **Colmekill** Iona, an island in the Hebrides
and a very early seat of Celtic Christian-
ity. As the text makes clear, it was the
traditional burial-place of Scottish kings
(not to be confused with St Columb's

Inch, 1.2.62, though the name also
derived from St Columba).
36–7 **Well ... Well ... well** The repetition
stresses the irony.
40 **benison** benediction, i.e. blessing
3.1.4 **stand in thy posterity** continue (*OED*
II.41) in your descendants

And set me up in hope? But hush, no more.   10
> *Sennet sounded.*
> *Enter Macbeth as King, Lady Macbeth as Queen,*
> *Lennox, Ross, Lords, and Attendants*

MACBETH
Here's our chief guest.
LADY MACBETH   If he had been forgotten,
It had been as a gap in our great feast,
And allthing unbecoming.
MACBETH (*to Banquo*)
Tonight we hold a solemn supper, sir,
And I'll request your presence.
BANQUO   Let your highness
Command upon me, to the which my duties
Are with a most indissoluble tie
Forever knit.
MACBETH
Ride you this afternoon?
BANQUO   Ay, my good lord.
MACBETH
We should have else desired your good advice,   20
Which still hath been both grave, and prosperous,
In this day's Council—but we'll take tomorrow.
Is't far you ride?
BANQUO
As far, my lord, as will fill up the time
'Twixt this and supper. Go not my horse the better,
I must become a borrower of the night
For a dark hour or twain.

10.2 *Macbeth as Queen,*] *not in* F   22 take] F; talk MALONE; take't KEIGHTLEY (*conj.* Warburton)

10.1 *Sennet* The word seems to occur only in stage directions of Elizabethan and Jacobean plays between 1590 and 1620; it evidently refers to a brief call on trumpet or cornet, and may be a corruption of the French 'signet', a 'small sign'.
Macbeth ... Queen F continues to refer to them as Macbeth and Lady throughout. F's 'Lady lenox' here is merely a misprint for 'Lady, Lenox'.
11–45 See Appendix A.
13 **allthing** everything, altogether
14 **solemn** formal, ceremonious

16 **Command upon me** i.e. lay your command upon me (cf. Leviticus 25: 21, 'Then I will command my blessing upon you ... and it shall bring forth fruit'); Banquo repudiates the proffered friendship of 'request'.
21 **still** continually
prosperous promoting prosperity
22 **take tomorrow** i.e. take it (Banquo's advice) tomorrow; when spoken, the words are indistinguishable from 'take't to-morrow'. Oxford accepts Malone's 'talk' which may be correct.

MACBETH   Fail not our feast.

BANQUO My lord, I will not.

MACBETH

We hear our bloody cousins are bestowed

In England and in Ireland, not confessing   30

Their cruel parricide, filling their hearers

With strange invention. But of that tomorrow,

When therewithal we shall have cause of state

Craving us jointly. Hie you to horse; adieu,

Till you return at night. Goes Fleance with you?

BANQUO

Ay, my good lord: our time does call upon's.

MACBETH

I wish your horses swift, and sure of foot—

And so I do commend you to their backs.

Farewell.   *Exit Banquo*

Let every man be master of his time   40

Till seven at night—

To make society the sweeter welcome,

We will keep ourself till supper-time alone—

While then, God be with you.

               Exeunt all but Macbeth and a Servant

Sirrah, a word with you: attend those men our

    pleasure?

SERVANT

They are, my lord, without the palace gate.

MACBETH

Bring them before us.   *Exit Servant*

             To be thus is nothing,

But to be safely thus. Our fears in Banquo

Stick deep, and in his royalty of nature

Reigns that which would be feared. 'Tis much he

    dares,   50

44.1] F (*Exeunt Lords.*)

33–4 **cause of state | Craving us jointly** state business calling urgently for our joint attention

44 **While** until

45–50 See Appendix A.

47–8 **To be ... thus** See Lady Macbeth's similar construction in 3.2.5–8.

48 **But to be** without being

49 **Stick deep** metaphorically like thorns in flesh

49–50 **royalty of nature | Reigns** i.e. 'king-like nature is the quality ...'. The words are metaphoric, but the allusion to the Weïrd Sisters' prophecy is apparent.

And to that dauntless temper of his mind
He hath a wisdom that doth guide his valour
To act in safety. There is none but he
Whose being I do fear; and under him
My genius is rebuked, as it is said
Mark Antony's was by Caesar. He chid the Sisters
When first they put the name of king upon me,
And bade them speak to him. Then prophet-like,
They hailed him father to a line of kings.
Upon my head they placed a fruitless crown,   60
And put a barren sceptre in my gripe,
Thence to be wrenched with an unlineal hand,
No son of mine succeeding; if't be so,
For Banquo's issue have I filed my mind,
For them the gracious Duncan have I murdered,
Put rancours in the vessel of my peace
Only for them, and mine eternal jewel
Given to the common enemy of man,
To make them kings, the seeds of Banquo kings.
Rather than so, come Fate into the list,   70
And champion me to th' utterance. Who's there?
      Enter Servant, and two Murderers

69 seeds] F; seed POPE

51 **to** in addition to
55 **genius** the attendant spirit in Roman mythology, Christianized as 'guardian angel'
56 **Mark Antony's ... Caesar** The allusion is to Octavius Caesar (Augustus); Shakespeare found it in Plutarch, and the soothsayer quotes it in *Antony* 2.3.16–20: 'Therefore, O Antony, stay not by his side. | Thy dæmon, that thy spirit which keeps thee, is | Noble, courageous, high, unmatchable, | Where Caesar's is not. But near him thy angel | Becomes afeard, as being o'erpowered.' (See Introduction, pp. 63–4.)
60–1 **fruitless ... barren** Alluding, of course, to their prediction that he would have no successors, but stressing childlessness, which they did not say.
61 **gripe** grasp, tenacious clutch; it can be merely an alternative for 'grip', but was used more intensively, which is appropriate here

64 **filed** 1. defiled, befouled; 2. sharpened (as for dagger)
65 **gracious** As well as its common meanings, 'gracious' was used of the deity, and so of outstanding piety.
66 **rancours** bitterness, figuratively poison **vessel** cup, here probably alluding to the communion chalice, see 1.7.11 'poisoned chalice'
67 **mine eternal jewel** my immortal soul
68 **common enemy of man** the Devil
69 **seeds** descendants; in this sense 'seed' is usual as a collective noun, but the plural here stresses the endless succession
70 **list** tilt-yard (for a tournament)
71 **champion** As a noun it usually means 'support'; as a verb it sometimes meant 'challenge' which would be more obvious here; but it is ambiguous whether Fate defies Macbeth or supports him. **utterance** last extremity (*OED* 2, from OF 'outrance'); i.e. death

Now go to the door, and stay there till we call.

Exit Servant

Was it not yesterday we spoke together?

MURDERERS It was, so please your highness.

MACBETH Well then, now have you considered of my
speeches: know that it was he, in the times past, which
held you so under fortune, which you thought had been
our innocent self. This I made good to you in our last
conference; passed in probation with you how you were
borne in hand, how crossed; the instruments; who   80
wrought with them—and all things else that might to
half a soul, and to a notion crazed, say 'Thus did
Banquo.'

FIRST MURDERER You made it known to us.

MACBETH I did so—and went further, which is now our
point of second meeting. Do you find your patience so
predominant in your nature, that you can let this go?
Are you so gospelled to pray for this good man, and for
his issue, whose heavy hand hath bowed you to the
grave and beggared yours for ever?   90

FIRST MURDERER We are men, my liege.

74, 115, 139 MURDERERS] F (*Murth.*); 1 *Mur.* STEEVENS   75 then, now] F; then now,
HUNTER   75–6 have you ... speeches:] F1; you have ... speeches; F3; you have ... speeches?
ROWE; have you ... speeches? POPE   78 self.] F; self? MUIR

74, 115, 139 MURDERERS Combined response is appropriate for these lines;
elsewhere F indicates which murderer is
speaking. Steevens's '1 Mur.' presumably indicates preference for a single
speaker over the common practice of
unison—many editors have followed
him.
75–91 See Appendix A.
75–6 **Well ... speeches** These words have
often been adjusted and repunctuated,
and frequently made into a question, but
the form 'now have you' is fairly common as a statement. Muir thought 'know
... fortune' was also a question, which
strains the syntax. I take 'know' to be
imperative.
76 **he** i.e. Banquo
79 **passed in probation** reviewed, in order to
prove
80 **borne in hand** Commonly glossed 'de-

ceived', as in *OED*, bear, *v.* I.3e, where it
is related to 'maintain a statement', i.e.
convince. But elsewhere in Shakespeare
it seems to be used in contexts suggesting
affection or support, though deceptive,
which contrasts more straightforwardly
with 'crossed'. See *Much Ado* 4.1.304–5,
Hamlet 2.2.66–7, *Cymbeline* 5.6.43;
Dent H94 gives 'to bear one in hand' as
proverbial, but without comment on its
meaning.
81 **wrought** worked, with a strong suggestion of 'moulded' (*OED*, wrought 1), i.e.
distorted.
81–2 **that might ... say** This seems to be
elaborated from the proverb 'he that has
but half an eye [sometimes "wit"] may
see it' (Dent H47).
82 **half a soul** half-wit
notion understanding, mind

MACBETH
    Ay, in the catalogue ye go for men,
    As hounds, and greyhounds, mongrels, spaniels, curs,
    Shoughs, water-rugs, and demi-wolves are clept
    All by the name of dogs; the valued file
    Distinguishes the swift, the slow, the subtle,
    The house-keeper, the hunter, every one
    According to the gift which bounteous nature
    Hath in him closed—whereby he does receive
    Particular addition from the bill   100
    That writes them all alike;—and so of men.
    Now, if you have a station in the file
    Not i'th' worst rank of manhood, say't,
    And I will put that business in your bosoms,
    Whose execution takes your enemy off,
    Grapples you to the heart and love of us
    Who wear our health but sickly in his life,
    Which in his death were perfect.
SECOND MURDERER   I am one, my liege,
    Whom the vile blows and buffets of the world
    Hath so incensed, that I am reckless what I do   110
    To spite the world.

94 clept] F (clipt)   106 heart ₍] POPE; heart; F

92–101 Muir quotes Erasmus, *Colloquia*
(trans. 1671), 482–3, comparing the
variety of dogs to the variety of men;
Erasmus refers to 'diverse Wolves, Dogs
of an unspeakable variety' with none of
Shakespeare's particularization. Nearly
200 pages later (661–2) the words 'genius' and 'bounteous nature' occur, but
they are not unusual; nor indeed is the
general proposition.
92 **catalogue** complete enumeration
94 **Shoughs, water-rugs, and demi-wolves**
OED is unhelpful about these beasts.
Johnson identified 'shough' with 'Shock',
a common name in the eighteenth century for long-haired pet dogs, especially
poodles, and Shakespeare's word might
have been pronounced 'shock'; water-
rugs are unknown, but presumably
rough-haired water dogs (?spaniels);
demi-wolf is presumably the same as
wolf-dog, said to be a cross between wolf
and dog, but possibly only a wolf-like
dog, e.g. Alsatian or German shepherd;
Scot said it was used for were-wolf, but

that would not be in this catalogue.
94 **clept** called. The word was common in
Middle English but already rare by 1600;
F's 'clipt' may also have the sense of
'embraced', and might even have been
suggested by the standard term for
grooming shaggy dogs.
95 **valued file** list of distinguishing qualities
97 **house-keeper** watch-dog
99 **closed** enclosed, limited to certain indi-
viduals
100 **addition** something added to the generic
name (see 1.3.106); combined here with
'bill' it seems to play with accounting,
deriving from 'valued file' in l. 95
from i.e. departing from; *OED* A.6, 'apart
or aside from', was used after words
which signify distinction, difference, etc.,
or else imply that the 'particular' is *from*
the generic ('all alike'), so the 'addition' is
from the 'bill'.
bill list, inventory (*OED*, *sb.*³, 5)
105 **takes your enemy off** removes, rids
(you) of (*OED* IX.58)

FIRST MURDERER   And I another,
So weary with disasters, tugged with fortune,
That I would set my life on any chance,
To mend it, or be rid on't.
MACBETH   Both of you
Know Banquo was your enemy.
MURDERERS   True, my lord.
MACBETH
So is he mine; and in such bloody distance,
That every minute of his being thrusts
Against my near'st of life; and though I could
With bare-faced power sweep him from my sight
And bid my will avouch it, yet I must not,   120
For certain friends that are both his and mine,
Whose loves I may not drop, but wail his fall
Who I myself struck down; and thence it is
That I to your assistance do make love,
Masking the business from the common eye
For sundry weighty reasons.
SECOND MURDERER   We shall, my lord,
Perform what you command us.
FIRST MURDERER   Though our lives—
MACBETH
Your spirits shine through you. Within this hour, at
      most,
I will advise you where to plant yourselves,
Acquaint you with the perfect spy o'th' time,   130
The moment on't, for't must be done tonight,
And something from the palace; always thought

---

112 **tugged with** mauled by (*OED*, tug 4b,
    which does not support Muir's sugges-
    tion of a metaphor from wrestling)
114 The proverb 'either mend or end', Dent
    M874.
116 **distance** 1. quarrel (from Old French
    'destance'); 2. remoteness (Old French
    'distance' from Latin 'distantia'), com-
    pare 'near'st' in l. 118.
118 **near'st of life** what most nearly affects
    my life, i.e. vitals
121 **For** because of
122 **but wail** i.e. but must wail ('must' is
    understood from 'may' before)

130 **perfect spy o'th' time** The general sense
    is clearly 'exact arrangements for place
    and time', but it is an odd phrase:
    presumably 'spy' is used in the senses
    usually confined to the verb, e.g. 'observe
    with hostile intent', 'keep watch' (as a
    noun 'spial' had this sense). Johnson
    thought it referred to the Third Murderer,
    who might then be a local guide, but the
    context makes this unlikely.
132 **something** somewhat, a sufficient dis-
    tance
    always thought 'it must be' is understood
    from l. 131.

That I require a clearness—and with him,
To leave no rubs nor botches in the work,
Fleance, his son, that keeps him company,
Whose absence is no less material to me
Than is his father's, must embrace the fate
Of that dark hour—resolve yourselves apart,
I'll come to you anon.
MURDERERS   We are resolved, my lord.
MACBETH
I'll call upon you straight: abide within,   140
It is concluded—Banquo, thy soul's flight,
If it find Heaven, must find it out tonight.   *Exeunt*

3.2   *Enter Lady Macbeth as Queen, and a Servant*
LADY MACBETH   Is Banquo gone from court?
SERVANT
Ay, madam, but returns again tonight.
LADY MACBETH
Say to the King, I would attend his leisure
For a few words.
SERVANT   Madam, I will.   *Exit*
LADY MACBETH   Nought's had, all's spent,
Where our desire is got without content;
'Tis safer to be that which we destroy,
Than by destruction dwell in doubtful joy.
        Enter Macbeth
How now, my lord, why do you keep alone,
Of sorriest fancies your companions making,   10
Using those thoughts which should indeed have died
With them they think on? Things without all remedy
Should be without regard—what's done, is done.

3.2.0 *Lady Macbeth as Queen*] F (*Lady*)

133 **clearness** Probably a play between
'clearance', i.e. distance from the castle,
and 'in the clear'; Holinshed wrote 'so
that … he might cleare himself'.
134 **rubs** roughnesses
botches bunglings
3.2.3 ff. The formality of her direction is
carried over into the courtly inversions of
ll. 10–12: she does not attempt more
familiar bluntness until l. 13.
5–6 **Nought's … content** Closely echoing

Macbeth in 3.1.47–8.
7–8 A paradox which Macbeth seems to
echo in ll. 21–4; 'safe' had the sense of
health, physical, mental, and spiritual
(theologically 'saved') as well as its more
usual one of 'free from danger'.
11 **Using** pursuing, i.e. allowing to become
customary
12–13 **Things … regard** Dent R71.1 cites
the sententious 'where there is no remedy it is folly to chide'.

MACBETH

We have scorched the snake, not killed it:
She'll close, and be herself, whilst our poor malice
Remains in danger of her former tooth.
But let the frame of things disjoint—
Both the worlds suffer—
Ere we will eat our meal in fear, and sleep
In the affliction of these terrible dreams   20
That shake us nightly.—Better be with the dead,
Whom we, to gain our peace, have sent to peace,
Than on the torture of the mind to lie
In restless ecstasy.
Duncan is in his grave:
After life's fitful fever he sleeps well,
Treason has done his worst; nor steel, nor poison,
Malice domestic, foreign levy, nothing,
Can touch him further.

LADY MACBETH   Come on—
Gentle my lord, sleek o'er your rugged looks,   30
Be bright and jovial among your guests tonight.

MACBETH

So shall I, love, and so I pray be you;
Let your remembrance apply to Banquo,
Present him eminence, both with eye and tongue—
Unsafe the while, that we must lave our honours
In these flattering streams, and make our faces
Vizards to our hearts, disguising what they are.

14 scorched] F (scorch'd); scotch'd THEOBALD

14 **scorched** slashed, as with a knife; older editions emended to 'scotched' because this sense had become obsolete

15 **close** 1. heal; 2. close with us, attack **malice** wickedness, harmfulness

16 **former tooth** i.e. previous intent to bite poisonously

17–18, 24–5 Both are commonly printed as single lines, see Appendix A.

17 **frame of things disjoint** i.e. structure of the universe fall apart; a similar metaphor is handled at length in Ulysses's speech, *Troilus* 1.3.82–123

18 **Both the worlds** the terrestrial and celestial spheres

21–4 **Better ... ecstasy** Compare ll. 7–8, and see Introduction, pp. 17–19.

24 **ecstasy** 'the state of being beside oneself with anxiety, astonishment, fear, or passion' (*OED*)

24–5 See ll. 17–18.

28 **levy** raising of men for war

33–4 The primary sense is 'pay attention to Banquo above the other guests', but Macbeth has already organized the murder and is directing her attention to Banquo as their greatest danger in the hope of securing her complicity—hence his elaboration of their danger in ll. 35–7.

35 **lave** wash

36 **flattering streams** 1. magically transforming waters; 2. streams of flattery (poured over Banquo)

37 **Vizards** masks (the word was used

LADY MACBETH You must leave this.

MACBETH

O, full of scorpions is my mind, dear wife—
Thou know'st that Banquo and his Fleance lives.   40

LADY MACBETH

But in them nature's copy's not eterne.

MACBETH

There's comfort yet, they are assailable;
Then be thou jocund—ere the bat hath flown
His cloistered flight, ere to black Hecate's summons
The shard-born beetle with his drowsy hums
Hath rung night's yawning peal, there shall be done
A deed of dreadful note.

LADY MACBETH   What's to be done?

MACBETH

Be innocent of the knowledge, dearest chuck,
Till thou applaud the deed—Come, seeling night,
Scarf up the tender eye of pitiful day,   50

specifically of prostitutes masking themselves in public)

41 1. their descendants won't go on for ever;
2. they are not immortal. Macbeth picks
up the second sense, while she retreats
into the first.
copy 1. pattern or mould from which
copies are made (*OED* IV.8c), hence the
copies themselves (descendants); 2. copi-
ousness, the original sense and commonly linked with 'Nature' (see 'bounteous nature' in 3.1.98). The traditional
explanation (*OED* I.5b, *fig.*) as a metaphor from the legal term 'copyhold' (a
limited form of freehold) seems to be quite
irrelevant: see Muir's refutation, based
on an article by Clarkson and Warren in
Modern Language Notes, 55, pp. 483–93.

43 **jocund** 1. expressing cheerfulness; 2.
being cheerful
bat Associated with night, monastic
ruins, and witchcraft.

44 **black Hecate** Hecate was not conventionally described as black; as moon-goddess
she was certainly associated with night,
but as goddess of the witches she seems to
have become black, as in black magic,
and so could be distinguished from 'white
Diana', the virgin goddess of the moon.

45 **shard-born** 1. born in dung (shard-

beetle); 2. borne on wings that resemble
broken pottery (shards). According to
OED 'shard' as a term for beetles' wings
was not used before 1775 when it was
derived from misunderstanding this line.

47 **What's to be done?** That she blinds
herself to the obvious only confirms Macbeth's awareness of her withdrawal. He
has begun his apostrophe to Night in
l. 43 which, though clearly delivered in
her hearing, is effectively soliloquy: an
invocation in the same mode as hers in
1.5.37–53 though here there is no suggestion of actual ritual.

48 **chuck** a term of endearment, equivalent
to 'chick'

49–56 Empson, *Seven Types*, pp. 18–20,
81–2, offered a celebrated analysis of
these lines as an example of the verbal
creation of a 'heavy atmosphere' out of
meanings that seem at first irrelevant. I
have used a few points in commentary
below, but the whole is too long and
densely argued for extraction: it has to be
read in its own context.

49 **seeling** blinding—a technical term for
closing the eyes of a hawk by sewing up
the lids and tying the thread behind the
head while training it for hawking

50 **Scarf up** tie up (as in seeling); compare
'blanket of the dark' in 1.5.52

And with thy bloody and invisible hand
Cancel and tear to pieces that great bond
Which keeps me pale. Light thickens,
And the crow makes wing to th' rooky wood;
Good things of day begin to droop and drowse,
Whiles night's black agents to their preys do rouse.
—Thou marvell'st at my words; but hold thee still,
Things bad begun make strong themselves by ill—
So prithee go with me.   *Exeunt*

3.3   *Enter three Murderers*
FIRST MURDERER (*to Third Murderer*)
But who did bid thee join with us?
THIRD MURDERER   Macbeth.
SECOND MURDERER (*to First Murderer*)
He needs not our mistrust, since he delivers
Our offices and what we have to do
To the direction just.
FIRST MURDERER (*to Third Murderer*) Then stand with us;
The west yet glimmers with some streaks of day.

---

51 **thy ... hand** Hunter suggests that this metaphor is from the falconer.

52 **that great bond** The legal metaphor may well derive from the latent pun seeling/ sealing, but the bond itself is not clearly identified. Steevens compared *Richard III* 4.4.77, 'Cancel his bond of life, dear God, I pray' and a similar use in *Cymbeline* 5.4.27–8, and so referred it to Banquo's life; Moberly thought it might be the bond of destiny announced by the Weïrd Sisters; Foakes that it was his obligation of love and loyalty to his subjects and other men generally; Hunter that it was the moral law. It seems to me that the phrase is so resonant that, though it may include many or all of these, it reaches beyond them to the whole structure of existence, recalling ll. 17–18; the nearest analogue in 1.5 is 'nature's mischief' in l. 49, but that is much less comprehensive. There may also be a play on 'bond' and 'bondage', the imprisonment in fear that keeps Macbeth pale.

53 **pale** 1. with fear; 2. with pity or cowardice; fenced in (Hunter)
Light thickens i.e. towards darkness; see

'Come, thick night', 1.5.49

54 **crow ... rooky wood** Empson (p. 19) commented 'Rooks live in a crowd and are mainly vegetarian; crow may be another name for a rook, especially when seen alone, or it may mean the solitary carrion crow', i.e. that Macbeth sees himself both as one with the rest, *and* as the scavenger that eats their flesh. In any case, all crows are black, and like beetles and bats associated with witchcraft. For other associations of 'rooky', see Empson, pp. 81–2.

58 Elaborating the sententious 'crimes are made secure by greater crimes', Dent C826.

3.3.0 *Enter three Murderers* Speculation about the identity of the Third Murderer (Macbeth himself, Destiny, etc.) is absurd: l. 2 fully explains his presence and is wholly in line with the dramatization of mistrust from 2.3 onwards (see also 3.4.132–3).

2 **He needs not our mistrust** i.e. we need not mistrust him (the Third Murderer)

4 **To the direction just** i.e. just as the directions were given us

Now spurs the lated traveller apace
To gain the timely inn, and near approaches
The subject of our watch.
THIRD MURDERER   Hark, I hear horses.
BANQUO (*within*)
Give us a light there, ho!
SECOND MURDERER   Then 'tis he—the rest,   10
That are within the note of expectation,
Already are i'th' court.
FIRST MURDERER   His horses go about.
THIRD MURDERER
Almost a mile—but he does usually,
So all men do, from hence to th' palace gate
Make it their walk.
     Enter Banquo and Fleance, with a torch
SECOND MURDERER A light, a light!
THIRD MURDERER 'Tis he.
FIRST MURDERER Stand to't.
BANQUO
It will be rain tonight.
FIRST MURDERER   Let it come down.
  ⌐*They attack Banquo. First Murderer strikes out the light*⌐
BANQUO O, treachery!   20
Fly good Fleance, fly, fly, fly—thou mayst revenge.
O slave!   ⌐*Dies. Fleance escapes*⌐
THIRD MURDERER
Who did strike out the light?
FIRST MURDERER   Was't not the way?
THIRD MURDERER
There's but one down: the son is fled.
SECOND MURDERER   We have lost
Best half of our affair.
FIRST MURDERER   Well, let's away,
And say how much is done.   *Exeunt*

3.3.8 and] F2; end F1   15.1] THEOBALD (*subst.*); *not in* F   22 *Dies … escapes*] POPE; *not in* F2

7 **lated** belated
11 **note of expectation** list of expected guests
15.1 F has no direction here: it is generally
    assumed that the First Murderer strikes

out the light because he defends the
action in l. 19.
26 **Exeunt** Presumably dragging Banquo's
body with them.

3.4 *Banquet prepared.*
    Enter King Macbeth, Lady Macbeth as Queen, Ross,
    Lennox, Lords, and Attendants

MACBETH You know your own degrees, sit down; at first
and last, the hearty welcome.

LORDS Thanks to your majesty.

MACBETH Ourself will mingle with society, and play the
humble host; our hostess keeps her state, but in best
time we will require her welcome.

LADY MACBETH Pronounce it for me sir, to all our friends,
for my heart speaks they are welcome.

    Enter First Murderer

MACBETH (*to Lady Macbeth*)
  See, they encounter thee with their hearts' thanks.
  Both sides are even—here I'll sit i'th' midst;   10
  Be large in mirth, anon we'll drink a measure
  The table round. (*To First Murderer*)
  There's blood upon thy face.

FIRST MURDERER   'Tis Banquo's then.

MACBETH
  'Tis better thee without than he within.
  Is he dispatched?

3.4.0.1 *King Macbeth, Lady Macbeth as Queen*] F (*Macbeth, Lady*)

3.4.0 Table(s) and chairs (or stools), and dishes, etc., have all to be brought on to the stage, together with thrones for the King and Queen. If the thrones were set in the discovery space at the back of the stage and revealed by drawing back the curtains, they would be suitably remote to justify l. 5. Presumably Macbeth first escorts the Queen to her throne, possibly both sit, but by l. 4 he has come forward in less formal hospitality, leaving her seated. Her reserve is doubtless a continuation of her bewilderment at the end of 3.2, only broken by the provocation of his distraction. 'Banquet' was used in its modern sense of a feast, from 1483 (*OED* 1), but could also mean sweetmeats, dessert, or snacks between meals. The formality of this scene seems to imply a feast.

1–8 F prints these lines as very erratic verse, see Appendix A. Verse patterns do become recognizable in ll. 9–12, and the rhythm becomes decisive in Macbeth's exchanges with the Murderer.

1 **degrees** i.e. formal order of precedence for state occasions

4 **society** company

5 **state** throne (or other chair of state)

6 **require** request

9 i.e. they toast the Queen

10 **i'th' midst** Either at the end, or more probably (since Macbeth is stressing informality) half-way down one side of the table. The table is often set across the stage with the vacant chair back to the audience, which is convenient since Macbeth never sits in it and Banquo's identity is effectively masked when he first occupies it.

11 **large** unrestrained

12 The incomplete line gives space for recognition, moves, etc.

14 **thee ... within** outside you than inside him

FIRST MURDERER　　My lord, his throat is cut,
　That I did for him.
MACBETH　　　　　　　Thou art the best o'th' cut-throats,
　Yet he's good that did the like for Fleance;
　If thou didst it, thou art the nonpareil.
FIRST MURDERER
　Most royal sir—
　Fleance is scaped.
MACBETH　　　　　　Then comes my fit again;　　　　　　20
　I had else been perfect—
　Whole as the marble, founded as the rock,
　As broad, and general, as the casing air;
　But now I am cabined, cribbed, confined, bound in
　To saucy doubts and fears.—But Banquo's safe?
FIRST MURDERER
　Ay, my good lord: safe in a ditch he bides,
　With twenty trenchèd gashes on his head,
　The least a death to nature.
MACBETH　　　　　　　　Thanks for that—
　There the grown serpent lies, the worm that's fled
　Hath nature that in time will venom breed,　　　　30
　No teeth for th' present.—Get thee gone, tomorrow
　We'll hear ourselves again.　　　*Exit First Murderer*
LADY MACBETH　　　　My royal lord,
　You do not give the cheer; the feast is sold

---

18 **nonpareil** one without equal ('the' be-
　cause there can be only one)
19–21 I have retained F's apparently eccen-
　tric arrangement of these lines because
　they cannot be made regular, and may
　indicate pauses for hesitation and shock;
　see Appendix A.
20 **fit** 'Ague fit of fear' occurred in proverbs,
　Dent A82.1.
22 'As hard as marble' and 'as fixed as rock'
　were both proverbial, Dent M638.1,
　R151.
　founded securely based
23 Dent A88 suggests elaboration of 'as free
　as air'.
　casing 1. surrounding; 2. enclosing (here
　paradoxical)
24 **cabined** shut up in a hut or hovel
　cribbed 'crib' was used of an ox-stall, a
　small room, or a cabin as above
25 **saucy** literally 'insolent towards su-

periors', hence these lower emotions dis-
turb the higher peace of mind
27 **trenchèd** cut (especially of severe wounds
　in flesh)
32 **We'll hear ourselves again** we'll hear
　each other again, i.e. talk further; there
　is, I think, an ambiguous allusion to the
　royal 'we', identifying king and murderer
33–7 Highly convoluted language of cour-
　tesy, extending a slight enough meaning
　from the sententious 'welcome is the best
　cheer' (Dent W258): to be enjoyed, the
　feast must be given (not sold)—otherwise
　it's better to eat at home. Away from
　home, meat needs the sauce of constant
　welcome. 'Meat' puns with 'meeting';
　'ceremony' had the sense (*OED* 2) of
　courtesy, politeness, but here it puns
　with 'cere', the wax used in laying out
　the dead.

That is not often vouched while 'tis a-making;
'Tis given, with welcome—to feed were best at home;
From thence, the sauce to meat is ceremony,
Meeting were bare without it.
    Enter the Ghost of Banquo, and sits in Macbeth's
    place
MACBETH   Sweet remembrancer—
Now good digestion wait on appetite,
And health on both.
LENNOX   May't please your highness sit?
MACBETH
Here had we now our country's honour roofed,   40
Were the graced person of our Banquo present,
Who may I rather challenge for unkindness
Than pity for mischance.
ROSS   His absence, sir,
Lays blame upon his promise. Please't your highness
To grace us with your royal company?
MACBETH
The table's full.
LENNOX   Here is a place reserved, sir.
MACBETH Where?
LENNOX
Here, my good lord. What is't that moves your
    highness?
MACBETH
Which of you have done this?
LORDS   What, my good lord?

34 a-making;] F (a making:); a-making, MALONE

37.1–2 *Enter the Ghost ... place* It should be
stressed that this direction is in F, making
it clear that an actual ghost did enter on
the Jacobean stage (see Introduction,
p. 4); its exact timing is not reliable,
because it probably derives from a marginal note, but the delay before Macbeth
sees it can be effective, and the audience's
attention is drawn to the device by l. 39.
   remembrancer one whose business it is to
   remind
40 **our country's honour** all our nobility (it
   seems odd that no reference is made to
   Macduff's absence)
41 **graced** endowed with grace

42–3 Another convoluted courtesy: 'whom
I may rather reproach for unkindness to
us, than sympathize with for his misfor-
tune in missing our feast'.
48 See Appendix A.
49–50 Initially, 'Which of you have done
this' appears to mean 'played this trick on
me', since the Lords ('you') have led him
to the seat; but it can also mean 'murdered Banquo', which is confirmed unambiguously by 'Thou canst not say I did
it'. The trick is thus established as theatrical and defines the precise nature of this
illusion (see Introduction, p. 4).

MACBETH (*to Banquo's Ghost*)
Thou canst not say I did it—never shake   50
Thy gory locks at me.

ROSS
Gentlemen rise, his highness is not well.

LADY MACBETH
Sit, worthy friends, my lord is often thus,
And hath been from his youth. Pray you keep seat,
The fit is momentary, upon a thought
He will again be well. If much you note him
You shall offend him and extend his passion;
Feed, and regard him not. (*Aside to Macbeth*) Are you a
    man?

MACBETH (*aside to Lady Macbeth*)
Ay, and a bold one, that dare look on that
Which might appal the Devil.

LADY MACBETH (*aside to Macbeth*) O proper stuff!   60
This is the very painting of your fear,
This is the air-drawn dagger which you said
Led you to Duncan. O these flaws and starts,
Impostors to true fear, would well become
A woman's story at a winter's fire,
Authorized by her grandam—shame itself,
Why do you make such faces? When all's done
You look but on a stool.

MACBETH
Prithee see there—behold, look, lo, how say you?
    (*To Banquo's Ghost*)
Why what care I, if thou canst nod, speak too.   70

55 **upon a thought** in a moment, as in the proverb 'as swift as thought', Dent T240.
56 **note** take notice of
57 **passion** overpowering emotion, affliction of body or mind, outburst
58 **Are you a man** A particular concern of the rest of the scene: 'man' is set against devil (60), woman (65), human (77), various birds and animals (73, 101f., 126), baby (107), etc.—and, of course, the Ghost.
60 **proper stuff** complete rubbish
61 **painting of your fear** 1. image of what you fear; 2. illusion created by your fear
62 **air-drawn dagger** 1. dagger delineated in air; 2. dagger moving through the air
63 **flaws** sudden passions (*OED* figurative use of 'flaw' *sb.*¹², meaning sudden blast of rough weather)
66 **Authorized** attributed to—1. as authority, 2. as author
    shame itself i.e. for very shame
67 **When all's done** Dent A211.1 gives this as a common phrase.
68 **stool** seat of any kind, including throne (see l. 83), but here used to belittle the stage furniture which is all she sees, while we, like Macbeth, see Banquo sitting on it

If charnel houses and our graves must send
Those that we bury back, our monuments
Shall be the maws of kites.   ⌜*Exit Ghost*⌝
LADY MACBETH What, quite unmanned in folly?
MACBETH
If I stand here, I saw him.
LADY MACBETH   Fie for shame.
MACBETH (*aside*)
Blood hath been shed ere now, i'th' olden time
Ere human statute purged the gentle weal;
Ay, and since too, murders have been performed
Too terrible for the ear. The times has been,
That when the brains were out the man would die,   80
And there an end; but now they rise again
With twenty mortal murders on their crowns,
And push us from our stools. This is more strange
Than such a murder is.
LADY MACBETH   My worthy lord,
Your noble friends do lack you.
MACBETH   I do forget:
Do not muse at me, my most worthy friends—
I have a strange infirmity, which is nothing
To those that know me. Come, love and health to all,

73 *Exit Ghost*] F2; *not in* F1   77 human] F (humane)   79 times has] F1; times have F2; time
has WHITE, OXFORD

71 **charnel houses** vaults or buildings for the bones of the dead
72–3 **our monuments ... kites** the only monuments (graves) we shall have will be the stomachs of birds of prey, as in Spenser, *Faerie Queene*, 2.8.16: 'What herce or steed (said he) should he have dight, | But be entombed in the raven or the kight?' More complicated explanations have been offered, based on the biblical idea that bodies could be given to birds to ensure their non-return, which is discussed by Scot in *Discovery of Witchcraft*, book 5, chapter 6, and may influence the image here. Kites are said to have a habit of disgorging undigested material, but the connection is tenuous.
73 **Exit Ghost** F2's placing of the Ghost's exit seems a reasonable guess, since Macbeth addresses him in l. 70 and refers to him in the past tense in l. 75.
75 See Dent S818 'as true as you stand there'.

77 **human** 1. man-made; 2. humane (F spells 'humane', which was usual for both senses)
    gentle weal the good of the commonwealth; Empson, *Seven Types*, p. 203, quotes this as a case of a word used in place of its opposite 'because the *weal* is conceived as "ungentle" before it was *purged* and *gentle* afterwards'
79 **times has** The common singular verb for a plural noun, but F2 already thought it worth correcting to 'times have'; although it may sound odd, 'times' is collective in sense and so might still govern 'has'.
81 **And there an end** Common in proverbs, Dent B458.
82 **twenty mortal murders** See l. 27, 'twenty trenchèd gashes'.
    crowns 1. heads; 2. kings
83 **stools** thrones (see l. 68)

Then I'll sit down—give me some wine, fill full:
    Enter Ghost
I drink to th' general joy o'th' whole table,   90
And to our dear friend Banquo, whom we miss;
Would he were here; to all, and him, we thirst,
And all to all.
LORDS   Our duties, and the pledge.
MACBETH (*to the Ghost*)
Avaunt, and quit my sight, let the earth hide thee—
Thy bones are marrowless, thy blood is cold;
Thou hast no speculation in those eyes
Which thou dost glare with.
LADY MACBETH   Think of this, good peers,
But as a thing of custom: 'tis no other,
Only it spoils the pleasure of the time.
MACBETH   What man dare, I dare: 100
Approach thou like the rugged Russian bear,
The armed rhinoceros, or th' Hyrcan tiger,
Take any shape but that, and my firm nerves
Shall never tremble. Or be alive again,
And dare me to the desert with thy sword—
If trembling I inhabit then, protest me
The baby of a girl. Hence, horrible shadow,
Unreal mock'ry, hence.   *Exit Ghost*
        Why so, being gone
I am a man again—pray you sit still.

108 *Exit Ghost*] ROWE; *not in* F

89.1 The Ghost's entry here invites the audience to see it before Macbeth does, and so stresses the irony of l. 91; but Macbeth may still see it before he actually addresses it.

92 **thirst** 1. want to drink; 2. long for (Banquo's company, and general amity in the toast)

96 **speculation** power of vision, especially comprehending vision

97–100 **Which thou dost ... dare** Macbeth's speech is not interrupted: his l. 100 and Lady Macbeth's 'Think ... peers' (l. 97) *both* complete the first half of l. 97; thus the effect is of simultaneous speech, though it will not actually be so.

102 **armed** i.e. by its horn and by its skin, represented (e.g. by Dürer) as if it were plate armour

102 **Hyrcan tiger** Pliny, *Natural History*, viii. 18, uses Hyrcanian sea for the Caspian, and the country south of it was Hyrcania, reputed to be heavily stocked with wild beasts; his book was widely read in grammar schools, and translated by Philemon Holland in 1601—the rhinoceros is mentioned close to the tiger.

106 **If trembling I inhabit** 1. if, trembling, I stay indoors; 2. if trembling inhabits me; see *As You Like It*, 3.3.7–8: 'O knowledge ill-inhabited; worse than Jove in a thatched house' which is similarly ambiguous

107 **baby of a girl** baby girl (as in 'fool of a man')

108 *Exit Ghost* F again leaves out the direction, which was added by Rowe; but here the dialogue leaves no doubt about its

LADY MACBETH

You have displaced the mirth, broke the good meeting,   110
With most admired disorder.

MACBETH   Can such things be,
And overcome us like a summer's cloud,
Without our special wonder? You make me strange
Even to the disposition that I owe,
When now I think you can behold such sights
And keep the natural ruby of your cheeks,
When mine is blanched with fear.

ROSS   What sights, my lord?

LADY MACBETH

I pray you speak not; he grows worse and worse,
Question enrages him—at once, good night.
Stand not upon the order of your going,   120
But go at once.

LENNOX   Good night, and better health
Attend his majesty.

LADY MACBETH   A kind good night to all.
                              Exeunt Lords and Attendants

MACBETH

It will have blood they say: blood will have blood.
Stones have been known to move, and trees to speak;
Augures, and understood relations, have
By maggot-pies, and choughs, and rooks brought forth

122.1] F (*Exit Lords.*)

placing, only about its means: Macbeth seems to pause so briefly that a long walk off stage would be awkward, and possibly a trap was used.

111 **admired** wondered at
112 **overcome** 1. cover (as a cloud does the sun); 2. overwhelm
113–14 **You ... owe** i.e. you make me feel alienated from my (male) nature
114 **owe** own
120 i.e. do not insist on due precedence
123 **blood will have blood** Proverb, Dent B458.
124 As disturbances of nature, and so as auguries; speaking trees are common in literature, at least since Virgil, but moving stones are less well documented: scholars have made vague references to

the tombs of murdered men opening themselves, or to prehistoric stones used as an ordeal, moving to an innocent touch but immobile to a guilty one—but no convincing analogues have been produced.

125 **Augures** disyllabic form of 'auguries', now obsolete; 'augurs' were diviners, 'auguries' the divinations
understood relations interpretation as augury of the relative placings of the entrails of sacrificed birds
126 **maggot-pies, and choughs, and rooks** All crows, used in sacrificial augury, and echoing 3.2.54; they are, of course, birds of ill omen.
maggot-pies magpies
choughs Used then of any of the smaller chattering species of crow.

The secret'st man of blood. What is the night?

LADY MACBETH
Almost at odds with morning, which is which.

MACBETH
How sayst thou that Macduff denies his person
At our great bidding?

LADY MACBETH   Did you send to him, sir?   130

MACBETH
I hear it by the way; but I will send;
There's not a one of them but in his house
I keep a servant fee'd. I will tomorrow—
And betimes I will—to the Weïrd Sisters.
More shall they speak: for now I am bent to know
By the worst means the worst; for mine own good,
All causes shall give way. I am in blood
Stepped in so far, that should I wade no more,
Returning were as tedious as go o'er.
Strange things I have in head, that will to hand,   140
Which must be acted, ere they may be scanned.

LADY MACBETH
You lack the season of all natures, sleep.

MACBETH
Come, we'll to sleep—my strange and self-abuse

---

134 Weïrd] F (weyard)   136 worst; ... good,] JOHNSON (*subst.*); worst, ... good, F

---

127 **secret'st man of blood** most successfully
concealed murderer
What is the night? What time of night
is it?

133 **fee'd** i.e. bribed as a spy

133-4 **I will ... Sisters** I have used dashes for
F's brackets, and so retained its sense;
Muir believes it should be repunctuated
so that 'I will tomorrow' refers to Macduff
and 'betimes I will' to the Weïrd Sisters—
thus allowing a vaguer gap in time before
4.1. I think this would be strained in
rhythm and hard to stress so that it was
clear, but it would be possible. The obvious sense is that he will go straightaway.

134 **betimes** early, speedily; in good time

136 **By the worst means** i.e. by witchcraft
for mine own good Goes equally with
both clauses, both before and after.

137 **All causes shall give way** 1. everything
else must wait; 2. all matters must give

way to my good

137-9 **I am in blood ... as go o'er** Dent refers
to two proverbs, 'having wet his feet he
cares not how deep he goes' (F565.1),
'over shoes, over boots' (S379).

139 i.e. it would be as tedious to retreat as to
go on forward

141 i.e. which must be done before they are
reflected on (or talked about)

142 **season** 1. time for a particular event (an
old sense); 2. seasoning, as preservative
(e.g. salt)

143-4 Dent refers to proverbs 'use makes
mastery' (U24), 'custom makes sin no
sin' (C934).

143 **my strange and self-abuse** my strange
self-abuse
self-abuse 1. self-deception (about the
ghost); 2. self-denigration (of his manhood)

Is the initiate fear, that wants hard use—
We are yet but young indeed.   *Exeunt*

3.5   *Thunder.*
        Enter the three Witches, meeting Hecate
FIRST WITCH
    Why how now Hecate, you look angerly?
HECATE
    Have I not reason, beldams as you are?
    Saucy, and overbold, how did you dare
    To trade and traffic with Macbeth
    In riddles, and affairs of death,
    And I the mistress of your charms,
    The close contriver of all harms,
    Was never called to bear my part
    Or show the glory of our art?
    And which is worse, all you have done   10
    Hath been but for a wayward son,
    Spiteful and wrathful who, as others do,
    Loves for his own ends, not for you.
    But make amends now: get you gone,
    And at the pit of Acheron
    Meet me i'th' morning; thither he
    Will come, to know his destiny.
    Your vessels, and your spells, provide,
    Your charms, and everything beside—
    I am for th' air: this night I'll spend   20
    Unto a dismal, and a fatal end;

145 indeed] F; in deed THEOBALD
    3.5.0.1 *Hecate*] F (*Hecat*)   2 are?] F; are, CAPELL   3 overbold,] F (ouer-bold,); overbold?
CAPELL

144 **the initiate fear** fear on first experience
145 **indeed** 1. in truth; 2. in such a deed
3.5 This scene is generally regarded as an
     addition, and Shakespeare's responsi-
     bility for it is often doubted; see Introduc-
     tion, pp. 51–5, 57–9.
2–3 Most editors repunctuate, but F is as
     good as any; the sense is not much
     changed anyhow.
2   **beldams** hags
7   **close** secret
11   **wayward** wrong-headed, perverse, not
     following others' wishes; Hecate recog-

nizes that he is not a potential convert
15   **pit of Acheron** Hell. Acheron was a river
     in Hell: 'pit' may suggest the association
     of witches with old quarries, gravel-
     pits, etc.
18   **vessels** utensils, jars, etc.
20   **th' air** the atmosphere within the moon's
     sphere, which she inhabits as moon-
     goddess
20–1 **this night ... end** i.e. preparing the
     dismal use of apparitions, 'fatal' both as
     'destined' and as 'destructive'

Great business must be wrought ere noon.
Upon the corner of the moon
There hangs a vap'rous drop, profound,
I'll catch it ere it come to ground;
And that distilled by magic sleights
Shall raise such artificial sprites,
As by the strength of their illusion
Shall draw him on to his confusion.
He shall spurn fate, scorn death, and bear   30
His hopes 'bove wisdom, grace, and fear:
And you all know, security
Is mortals' chiefest enemy.
        Music.
          ⌈*A cloud appears*⌉
Hark, I am called; my little spirit see
Sits in a foggy cloud, and stays for me.
        Spirits sing within
FIRST SPIRIT
Come away, come away, Hecate, Hecate—
O come away.

33.1] This edition; *Musicke, and a Song* F, 1673; *Musick and Song.* | Heccate, Heccate, Heccate! O come away YALE, 1674; SPIRITS (*singing dispersedly within*) *and first two lines of song* OXFORD
35.1] This edition; *Sing within. Come away, come away, &c.* F, 1673; *Machine descends* YALE; *Sing within Machine descends* 1674.'; *The Song* OXFORD   36–68] BULL, DREXEL, WITCH, YALE, 1673, 1674, OXFORD; *not in* F (*no s.d.'s or prefixes in Bull or Drexel*)   36 FIRST SPIRIT] YALE (1 Sing w^th^in), 1673 (1.); Song: WITCH (*in* y^e *aire bracketed after* 36–7) *not in* 1674; SPIRITS *within* OXFORD   come away,] BULL; *not in* YALE, 1673, 1674   37 O] BULL; *not in* WITCH, OXFORD

24 **vap'rous drop** Steevens quoted Lucan, *Pharsalia*, vi. 669, on the *virus lunare*, a foam dropped by the moon in response to appropriate charms: it touched herbs or other objects and could transform them. **profound** i.e. with profound qualities (Johnson)
32–3 See Dent W152, 'the way to be safe is never to be secure'.
32 **security** confidence of safety
33.1 *Music* F calls for *Musicke, and a Song*, and then repeats the direction to sing after l. 35; I assume that there is only one song and that it accompanies Hecate's exit, but instrumental music may well begin here. The duplication could easily derive from marginal notes. Davenant's text interpolates a line similar to the opening of the song, and Oxford assumes that ll. 36–7 were sung as an anticipation of the full song. This is possible, but

Davenant's line does not fit the music and I suggest it was supplied as a spoken cue for l. 34; music alone will be a sufficient cue for Hecate, and F's direction is a general heading which does not require the song to start at once.
    The music might be provided simply by the consort of hautboys already required by F directions, but the addition here may have been accompanied by a larger band.
35.1 *within* Presumably behind the balcony above the stage.
36–68 The text of this song, not in F, is in three early seventeenth-century MSS, one of them Middleton's play, *The Witch*, and in three Restoration texts; it was edited in *Macbeth* for the first time in Oxford, in 1986. All seven texts are collated here for words, but Bull and Drexel are arranged for solo singer and so are not involved in s.d.'s and speech prefixes; texts, in

HECATE

I come, I come, I come, I come,
With all the speed I may,
With all the speed I may.
Where's Stadling? 40

SECOND SPIRIT Here.

HECATE Where's Puckle?

THIRD SPIRIT Here—

And Hopper too, and Hellway too;
We lack but you, we lack but you:
Come away, make up the count.

HECATE

I will but 'noint and then I mount,
I will but 'noint and then I mount.

A Spirit like a cat descends.
The other three Spirits appear above

FIRST SPIRIT

Here comes one to fetch his dues:

38 I come... I come] BULL; *twice only* YALE, 1674; *twice here, twice after l. 39* 1673 40] BULL; *not in* YALE 41] *no prefix* WITCH; 1. 1673 Stadling] BULL; Stadlin DREXEL, WITCH, OXFORD SECOND SPIRIT] YALE (2.); 3. 1673; SPIRIT ⌜*within*⌝ OXFORD; in yᵉ aire (*after* Here.) WITCH 42 HECATE] YALE; *not in* WITCH; 1. 1673 THIRD SPIRIT] YALE (3.); *not in* WITCH; 4. 1673; ANOTHER SPIRIT *within* OXFORD Here—] WITCH (heere∧), YALE (Here,), 1673 (Here;), 1674 (Here,), OXFORD (Here.) 43] *no prefix* WITCH (*brackets* 43–5 *in* yᵉ aire); *as one line with* Here YALE, 1673, 1674; OTHER SPIRITS *within* OXFORD Hopper] BULL (Hope), YALE (Hopper); Hoppo DREXEL, WITCH, OXFORD Hellway] BULL; Hellwaine WITCH, OXFORD 44] *no prefix* WITCH; *prefix* 1. YALE, 1673, 1674 lack ... lack] BULL; want ... want YALE, 1673, 1674 46 HECATE] WITCH; *not in* 1673 47] BULL; *not in* WITCH, OXFORD 47.1 A Spirit ... descends] WITCH (*before* 49–50); *not in* YALE, 1673, 1674 47.2 The other three Spirits appear above] OXFORD (subst.); above WITCH (*bracketed after* 48–50); *not in* YALE, 1673, 1674 48 FIRST SPIRIT] YALE (1.); *not in* WITCH; SPIRITS *above* OXFORD Here comes one] BULL; Heare comes one *down* DREXEL; Ther's one comes down WITCH, OXFORD; Here comes down one YALE, 1674; Here comes one, it is 1673 48–9 dues: | A kiss] BULL (subst.); Due, a kiss | YALE, 1673, 1674

either category, not listed against a particular item agree with copy. For full discussion, and for the early setting by Robert Johnson, see Appendix B, where problems of staging are also discussed.

41 **Stadling** Possibly from 'staddle', the bottom of a haystack.

42 **Puckle** Presumably connected with the Puck, Robin Goodfellow, the capricious, frequently malicious, elf of e.g. *A Midsummer Night's Dream*.

43 **Hopper** Possibly from 'hopper', a hop-picker; or it might be connected with the game of hopscotch.
Hellway Presumably implying a diabolic spirit.

46, 47 **'noint** anoint; presumably a ritual preparation for flight

47.1 *A Spirit ... descends* The Globe had a winch in the overhanging attic of the tiring-house; the New Globe, built in 1613, probably had more elaborate machinery but it is unlikely that all the spirits could have flown, either there or at the Blackfriars. One 'car' is enough for Cat and Hecate.

47.2 *The other ... above* The Witch changes its marginal directions here from 'in the air' to 'above'; this may be mere chance, but an entry on the balcony at this point fits the text, and it seems unlikely they remained invisible throughout.

48 **one** to 'down' is not required by the

A kiss, a coll, a sip of blood;
And why thou stay'st so long, I muse,

CAT   —I muse.   50

FIRST SPIRIT
  Since the air's so sweet and good.

HECATE
  O art thou come: what news?

CAT   —What news?

SECOND SPIRIT
  All goes still to our delight,
  Either come or else refuse.

CAT   —Refuse.

HECATE
  Now I am furnished for the flight,
  Now I go and now I fly,
  Malkin my sweet spirit and I.
        Hecate and the Cat ascend

THIRD SPIRIT
  O what a dainty pleasure's this

---

49 coll] BULL; cull DREXEL, 1673   a sip] BULL; sip 1673   50 stay'st] BULL; stayest DREXEL
I muse, | CAT I muse] This edition; I muse, I muse BULL, DREXEL, WITCH (*as new line*), OXFORD;
I muse YALE, 1673, 1674   51 sweet] BULL; freshe DREXEL   52 HECATE] WITCH; *not in* YALE,
1673; 2. 1674   what news? | CAT—What news?] This edition; what nues what nues BULL,
DREXEL, WITCH (*as new line*), OXFORD; what News? YALE, 1673, 1674   53 SECOND SPIRIT] YALE
(2); *not in* WITCH, 1674; SPIRIT LIKE A CAT OXFORD   still to] BULL; well to DREXEL; for YALE; fair
for 1673, 1674   54 refuse. | CAT—Refuse.] This edition; refuse refuse BULL, DREXEL, WITCH (*as
new line*), OXFORD; refuse YALE, 1673, 1674   55 HECATE] WITCH; *not in* YALE, 1673,
1674   Now] DREXEL; No BULL   I am] BULL; I'm 1674   *after* 55 WITCH *adds* Fire ⟨stone⟩:
hark, hark, the Catt sings a brave *Treble* in her owne language. *Not in* BULL, DREXEL, YALE, 1673,
1674, OXFORD   56 and] BULL; oh DREXEL; *not in* WITCH, 1673, OXFORD   57 Malkin] BULL;
Malking 1674   *Hecate and the Cat ascend*] WITCH (Hec. going up *before* 56), OXFORD (*subst.*); *not
in* YALE, 1673, 1674   58 THIRD SPIRIT] YALE (3:); *not in* WITCH, SPIRITS *and* HECATE
OXFORD   pleasure's this] BULL; pleasure is this DREXEL; pleasure 'tis WITCH, OXFORD

---

music, but it may be carelessly omitted
in Bull.

49 **coll** hug (*OED* has it as a noun here only
(*sb.*[1]), derived from the verb)

50 CAT **I muse** In *The Witch* Hecate's clown-
ish son Firestone comments after l. 55
'hark, hark, the Cat sings a brave treble
in her own language'; evidently she
mews, which explains the rhyming
echoes here, and in ll. 52 and 54. Firestone has no part in *Macbeth*, and his

intervention reduces a mildly comic effect
to the burlesque, characteristic of Middleton's Hecate; Davenant, if not someone
before him, seems to have thought it out
of place in *Macbeth*.

57 **Malkin** A familiar name for a cat (cf.
Graymalkin, 1.1.8), also used for women
in derogatory senses (slut, whore, etc.),
and for a female demon.

58 THIRD SPIRIT Davenant's arrangement is
logical: the Spirits sing, as the Weïrd
Sisters speak, in turn.

To ride in the air when the moon shines fair,
And feast and sing and toy and kiss.   60
CHORUS OF SPIRITS
Over woods, high rocks, and mountains,
Over seas, and misty fountains,
Over steeples, towers, and turrets,
We fly by night 'mongst troops of spirits.

                Hecate and the Cat disappear above

No ring of bells to our ears sounds,
No noise of wolves, or yelps of hounds,
No, nor the noise of waters' breach,
Or cannon's throat our height can reach.

                    Exeunt Spirits

FIRST WITCH
Come, let's make haste, she'll soon be back
    again.   *Exeunt*

3.6   *Enter Lennox, and another Lord*

LENNOX
My former speeches have but hit your thoughts

59] *as one line* YALE; *as two lines* WITCH, OXFORD   ride in the air when BULL; saile i'th'aire while YALE, 1673, 1674   60 BULL; And sing, and daunce, and toy, and kiss WITCH, OXFORD; To sing, to Toy, to Dance, and Kiss YALE, 1674; To sing, to toy and kiss 1673   61 CHORUS OF SPIRITS] This edition; *not in* WITCH, YALE, 1673, 1674, OXFORD   62 seas, and misty] BULL; seas ⟨and⟩ *(deleted)* cristell | mistris *(bracketed)* DREXEL; Seas, our Mistris WITCH, OXFORD; Hills and misty YALE, 1674; misty Hills and 1673   63 steeples] BULL; Steepe WITCH   turrets] DREXEL; tirritts BULL   64.1 *Hecate ... above*] This edition; *not in* WITCH, YALE, 1673, 1674, OXFORD   65] *no prefix* WITCH; *prefix* Cho. 1673   66 noise] BULL; howls DREXEL, WITCH, YALE, 1673, 1674, OXFORD   or] BULL; nor DREXEL, YALE, 1673, 1674; no WITCH, OXFORD   yelps] DREXEL; elps BULL   67 nor] BULL; not WITCH, OXFORD   68 Or] BULL; Nor DREXEL, YALE, 1673, 1674   cannon's throat] BULL (cannons), WITCH (cannons); cannons throats DREXEL, YALE, 1673, 1674; cannons' throat OXFORD   *Exeunt Spirits*] This edition; *not in* WITCH, YALE, 1673, 1674; *Exeunt into the heavens the Spirit like a Cat and Hecate* OXFORD   65–8] BULL; *after 68* No Ring of Bells &c. above} WITCH; *lines repeated, prefix* SPIRITS *above* OXFORD

61 CHORUS OF SPIRITS The tune changes here, but there is a more dramatic change at l. 64, from 4:4 time to 3:4, so 1673 may be right to indicate *chorus* there. I take both quatrains to be choral, the second after Hecate's disappearance.

62 For the textual confusion, see Appendix B.

63 turrets Bull's 'tirritts' makes a perfect rhyme.

64.1 *Hecate ... disappear above* See note on l. 61; the change of rhythm suggests some difference on stage, but it is a production question when precisely He-

cate disappears, so long as she is safely off before l. 69.

66 noise 'howls' is tempting, but the dissonance with 'hounds' may explain why an author avoided the cliché which a copyist supplied. Misreading is unlikely.

3.6.1–2 An indication of the oblique language essential when you can trust nobody in case your conversation (or your telephone) is tapped by a tyrannical government.

1 hit agreed with (*OED* II.16), coincided with

165

Which can interpret further; only I say
Things have been strangely borne. The gracious
    Duncan
Was pitied of Macbeth—marry he was dead;
And the right valiant Banquo walked too late,
Whom you may say, if 't please you, Fleance killed,
For Fleance fled—men must not walk too late.
Who cannot want the thought, how monstrous
It was for Malcolm and for Donalbain
To kill their gracious father? Damnèd fact,   10
How it did grieve Macbeth! Did he not straight
In pious rage the two delinquents tear,
That were the slaves of drink, and thralls of sleep?
Was not that nobly done? Ay, and wisely too—
For 'twould have angered any heart alive
To hear the men deny 't. So that I say,
He has borne all things well, and I do think,
That had he Duncan's sons under his key—
As, an 't please Heaven, he shall not—they should find
What 'twere to kill a father: so should Fleance.   20
But peace—for from broad words, and 'cause he
    failed
His presence at the tyrant's feast, I hear
Macduff lives in disgrace. Sir, can you tell
Where he bestows himself?

LORD   The son of Duncan,
From whom this tyrant holds the due of birth,
Lives in the English court, and is received
Of the most pious Edward, with such grace
That the malevolence of fortune nothing

3.6.24 son] THEOBALD; Sonnes F

8 **cannot want the thought** Empson, *Seven Types*, p. 209, comments that 'can want' is the meaning which comes through powerfully, though the '-not' 'acts as a sly touch of disorder'. A nice point, not invalidated by the fact that in Elizabethan English double negatives normally intensified rather than cancelled out, and 'want' (=lacks) is a negative verb.
10 **fact** 1. deed; 2. fact (if true)
12 **pious** faithful in appropriate regard for superiors, parents, etc.
19 **an't** if it; 'and' was standardly used in this sense
21 **broad** outspoken
24 **son** F's 'sonnes' is either a misprint or an author's slip—Donalbain went to Ireland and never re-enters the play.
27 **most pious Edward** Edward the Confessor; 'pious' here is in its modern sense, Edward was regarded as a saint
28–9 **That ... respect** i.e. his exile is not counted against him

Takes from his high respect. Thither Macduff
Is gone, to pray the holy King upon his aid   30
To wake Northumberland and warlike Seyward,
That by the help of these—with Him above
To ratify the work—we may again
Give to our tables meat, sleep to our nights,
Free from our feasts and banquets bloody knives,
Do faithful homage, and receive free honours—
All which we pine for now. And this report
Hath so exasperate their King, that he
Prepares for some attempt of war.

LENNOX Sent he to Macduff?   40

LORD
He did—and with an absolute 'Sir, not I',
The cloudy messenger turns me his back,
And hums; as who should say, 'You'll rue the time
That clogs me with this answer.'

LENNOX   And that well might
Advise him to a caution, t'hold what distance
His wisdom can provide. Some holy angel
Fly to the court of England, and unfold
His message ere he come, that a swift blessing
May soon return to this our suffering country,
Under a hand accursed.   50

LORD I'll send my prayers with him.   *Exeunt*

4.1   *Thunder.*
        Enter the three Witches

FIRST WITCH
Thrice the brinded cat hath mewed.

SECOND WITCH
Thrice, and once the hedge-pig whined.

---

38 their] F; the HANMER

30 **upon his aid** on his behalf
35 i.e. enjoy our feasts free from the threat of
   bloody knives
38 **exasperate** exasperated
   their King i.e. Edward. See Introduction,
   pp. 51–3, where the textual problems of
   this passage are discussed.
41 **absolute** peremptory (of Macduff's tone)
44 **clogs** obstructs
49–50 **suffering country, | Under** The syn-

tax is double: 'suffering' qualifies 'country' and also governs 'Under', i.e. 'suffering under a hand accursed'.

4.1.1 **brinded** strictly 'tawny with bars of
   another colour'; doubtless a sort of tabby
   cat presumably Graymalkin, the First
   Witch's familiar (1.1.8)
2 **hedge-pig** hedgehog; *OED* records the
   word only in this context, where it may
   be intended to imply the female.

THIRD WITCH
 Harpier cries—'tis time, 'tis time.
FIRST WITCH
 Round about the cauldron go—
 In the poisoned entrails throw.
 Toad, that under cold stone
 Days and nights has thirty-one,
 Sweltered venom sleeping got:
 Boil thou first i'th' charmèd pot.
ALL
 Double, double, toil and trouble,   10
 Fire burn and cauldron bubble.
SECOND WITCH
 Fillet of a fenny snake
 In the cauldron boil and bake;
 Eye of newt, and toe of frog,
 Wool of bat, and tongue of dog;
 Adder's fork, and blind-worm's sting,
 Lizard's leg, and howlet's wing:
 For a charm of powerful trouble,
 Like a hell-broth, boil and bubble.
ALL
 Double, double, toil and trouble,   20
 Fire burn and cauldron bubble.
THIRD WITCH
 Scale of dragon, tooth of wolf,

4.1.5 throw.] ROWE; throw ∧ F

3 **Harpier** The Third Witch's familiar; the
name has been associated with 'harpy',
the mythical bird-woman, spirit of vengeance, but there is no evidence for this:
Thomas lists only common animals and
insects as familiars.

6–8 **Toad** They were regarded as particularly ugly and loathsome; they do secrete
an acrid poison in the sweat glands
which is unpleasant rather than dangerous. Witches were often said to transform
themselves into toads, and to use their
poison. Foakes cites Agnes Tompson's
'confession' of her attempt to poison King
James in Edinburgh: 'took a black toad,
and did hang up the same by the heels,
three days, and collected and gathered
the venom as it dropped' *News from*

Scotland (1591), B2ᵛ (see Introduction,
p. 78 n. 1). Toads do, like other amphib-
ians, hibernate, but not in order to
sweat; why 'thirty-one days' (a calendar
month) nobody seems to know.

8 **Sweltered** *OED* has 'exuded like sweat',
but it is literally 'sweated out'.

12 **Fillet** 1. band, strip; 2. slice (as of meat or
fish); (2) is usually given, but (1) seems to
me at least as likely
fenny muddy (or, possibly, a fen-dwelling
snake)

16 **blind-worm** usually the slow-worm,
which has very small eyes, but also used
of the adder. Both slow-worms and newts
were thought, wrongly, to be poisonous.

17 **howlet** either owl, or young owl, or Little
Owl; a dialect form of owlet (*OED*)

Witch's mummy, maw and gulf
Of the ravined salt-sea shark;
Root of hemlock, digged i'th' dark;
Liver of blaspheming Jew,
Gall of goat, and slips of yew
Slivered in the moon's eclipse;
Nose of Turk, and Tartar's lips;
Finger of birth-strangled babe   30
Ditch-delivered by a drab:
Make the gruel thick and slab;
Add thereto a tiger's chawdron,
For th' ingredience of our cauldron.

ALL
Double, double, toil and trouble,
Fire burn and cauldron bubble.

SECOND WITCH
Cool it with a baboon's blood,
Then the charm is firm and good.
Enter Hecate, and the other three Witches

23 Witch's] F (Witches), SINGER; Witches' THEOBALD   34 cauldron] F (Cawdron)

23 **mummy** The dried and/or embalmed
flesh of a dead body was widely used for
medicinal purposes, by no means exclusively in witchcraft; hence the word was
used of powder or ointment derived from
the mummified flesh.
maw and gulf Both words refer to the
stomach and are connected with vora-
cious appetites; 'maw' can also be the
throat or gullet.
24 **ravined** glutted on its prey (cf. 'raven-
ous')
25 **hemlock** a common, poisonous, plant
dark Muir comments that the time when
herbs were gathered was thought to
affect their power.
26 **blaspheming** i.e. anti-Christian
27 **slips of yew** twigs of yew, used for mourn-
ing, but generally regarded (especially
the berries) as poisonous
28 **Slivered** cut or torn off
moon's eclipse seen as a particularly
sinister form of darkness (see l. 25)
29 **Turk, and Tartar** infidels, notorious for
cruelty and sensuality
30 **birth-strangled babe** As well as being a
horrid image, this refers to the idea that
unbaptized babies were condemned to
Hell.

31 **Ditch-delivered** delivered in a ditch—
suppressing the unwanted child
drab prostitute
32 **slab** Adjective from the noun (*OED*, *sb.*²2)
meaning slimy matter, ooze, or sludge;
later use of the adjective for 'semi-solid'
derives from this line.
33 **chawdron** entrails (especially as used in
cooking)
34 **ingredience** ingredients (for a mixture)
cauldron F spells it 'cawdron' here, but
'cauldron' elsewhere; either a compositor's slip from 'chawdron' above, or a
deliberate variant to stress the rhyme.
37 **baboon** Often stressed on the first syllable. Supposedly a lustful ape, i.e. 'hot',
related to 'cool' as the toad's 'cold stone'
was to 'swelter'.
38.1 The entry of Hecate is generally taken
to be an addition (as in 3.5); I take it to be
part of a revision affecting the whole
scene up to l. 146; see Introduction,
pp. 53–4.
the other three Witches There has been
much argument about this, largely un-
necessarily; clearly other witches are
required as a singing and dancing chorus
for Hecate (the Weïrd Sisters never do
either). My only doubt is whether 'three'

HECATE

O well done: I commend your pains,
And everyone shall share i'th' gains—   40
And now about the cauldron sing
Like elves and fairies in a ring,
Enchanting all that you put in.
        Music, and a song
FOURTH WITCH
Black spirits and white, red spirits and gray,
Mingle, mingle, mingle, you that mingle may.
Titty, Tiffin, keep it stiff in,
Firedrake, Puckey, make it lucky,
Liard, Robin, you must bob in.

43.1] F (*continuing 'Blacke Spirits, &c.'*); A Charme Song: about a Vessel WITCH   44–58]
WITCH, YALE, 1674, OXFORD; *not in* F, 1673   44 FOURTH WITCH] This edition; *not in* WITCH;
Hec: YALE, 1674, OXFORD   46] *no prefix* WITCH; *prefix* 1 Witch YALE, 1674; FOURTH WITCH
OXFORD   Titty, Tiffin] WITCH; Tiffin Tiffin YALE, 1674   47] *no prefix* WITCH; *prefix* 2: YALE
Firedrake,] WITCH, Fire drake∧ YALE, 1674   48] *no prefix* WITCH; *prefix* Hec: YALE
Liard,] WITCH; Lyer∧ YALE, 1674

is right: it is, of course, the magic number
for Hecate, but could so easily be repeated
accidentally from the usual s.d. for the
Witches' entries. In any case the song
and dance team soon increased to a full
chorus, see Introduction, pp. 34–47;
they may possibly have been children
originally, see l. 42.

42 Highly inappropriate for the Weïrd
Sisters, but perfectly possible for attendants on Hecate.

43.1–58 Middleton headed this 'A Charm
Song', appropriately; it may even have
been a traditional one, see Appendix B.
The only pre-Restoration text surviving is
in *The Witch*, and it is possible that it
always differed in *Macbeth*, but the differences in Davenant are slight (see notes on
ll. 54 and 56). The texts collated are
Witch, 1674, Yale, Oxford (it is not in
1673); as in 3.5.36–68, texts not cited
against a particular item agree with copy.
The brewing duplicates much of what the
Weïrd Sisters have already done, but here
the tone is specifically bawdy ('keep it stiff
in' etc.), as theirs was not. The 'charm' is
aimed in l. 55 at seducing a 'younker',
and in *The Witch* that is Hecate's intention; but the game seems more appropriate to the children who sing it than to
Hecate, even when she is (in Middleton) a
randy old witch. The term is even more

ironic in *Macbeth*, and Davenant (or
someone before him) made a clumsy
substitution. I have arranged the song for
the three other Witches, leaving Hecate
as spectator (see Appendix B).

46–8 These lines might well be given to
separate voices, e.g. Fifth, Sixth, Fourth
Witches, to keep sequence; Hecate's participation seems unlikely. The Spirits'
names are set out (with others) in a table
at the end of a pamphlet, which is quoted
in Appendix B, and referred to here as
Record; Scot summarized in an adden-
dum. Davenant reduced them to three by
treating the first name in each line as an
adjective.

46 Titty May derive from the extra teat
witches used to feed their familiars. 'Tet-
tey a he like a gray cat' (*Record*).
Tiffin 'Tiffing' was used of drinking.
'Tyffyn a she, like a white lamb' (*Record*).

47 Firedrake A dragon in German mytho-
logy; used of Bardolph in *Henry V*, and of
Face as the Alchemist's assistant in Ben
Jonson's play, *The Alchemist*. Not men-
tioned in *Record* or Scot.
Puckey See note on 3.5.42. Possibly
confused with Pygin in *Record* (Pidgin in
Scot), 'a she, like a black toad'.

48 Liard The word meant 'grey', probably
associated here with 'liar'. 'Lyard red like
a lion or hare' (*Record*).
Robin Goodfellow, see note on 3.5.42.

CHORUS OF WITCHES
  Round, around, around, about, about,
  All ill come running in, all good keep out.   50
FIFTH WITCH
  Here's the blood of a bat.
FOURTH WITCH   Put in that, O, put in that!
SIXTH WITCH
  Here's lizard's brain.
FOURTH WITCH   Put in a grain!
FIFTH WITCH
  The juice of toad, the oil of adder,
  Those will make the younker madder!
FOURTH WITCH
  Put in, there's all, and rid the stench.
SIXTH WITCH
  Nay, here's three ounces of the red-haired wench.
CHORUS OF WITCHES
  Round, around, around, about, about,
  All ill come running in, all good keep out.

49 CHORUS OF WITCHES] YALE (Chor:); *not in* WITCH; ALL OXFORD   Round, around, around]
WITCH; A round a round a round YALE; A round, a round 1674   51 FIFTH WITCH] This
edition; 1. witch WITCH, YALE, 1674; FOURTH WITCH OXFORD   FOURTH WITCH] This edition; Hec.
WITCH, YALE, 1674, OXFORD   O] WITCH; *before first* 'put' YALE, 1674   52 SIXTH WITCH] This
edition; 2. WITCH, YALE, 1674; FIFTH WITCH OXFORD   lizard's brain] YALE; Libbards Bane
WITCH, OXFORD (leopard's)   FOURTH WITCH] This edition; Hec. WITCH, YALE, 1674, OXFORD
a grain] YALE; againe WITCH   53 FIFTH WITCH] This edition; 1. WITCH, YALE, 1674;
FOURTH WITCH OXFORD   The ... the] WITCH; here's ... here's YALE, 1674   54] *no prefix* YALE;
prefix 2. WITCH; FIFTH WITCH OXFORD   younker] WITCH (yonker); charme grow YALE, 1674
55 FOURTH WITCH] This edition; Hec. WITCH, OXFORD; 2: YALE, 1674   Put ... stench] WITCH
(Put in: ther's all. and rid the Stench); Put in all these, 'twill raise the stanch YALE, 1674   56
SIXTH WITCH] This edition; Fire ⟨stone⟩ WITCH; Hec: YALE, 1674; A WITCH OXFORD   the]
WITCH; a YALE, 1674, OXFORD   57 CHORUS OF WITCHES] YALE (Chor:); all WITCH, OX-
FORD   57–8] WITCH (Round: around: around &c.); A round a round &c YALE, 1674

'Two spirits like toads, their names Tom
& Robin' (*Record*).

52 **lizard's brain ... a grain** *Witch* may be
  correct here, but there is no obvious
  reason for Davenant to introduce the
  lizard, whereas Crane could have misread; 'a grain' is stronger than 'again'.
  Lizards, like snakes, are often associated
  with witchcraft; see Appendix B. 'Grain',
  originally a measure of corn, was used as
  a minimal measure for any substance,
  especially solids; the lizard might be
  pulverized, but its dried brain would

hardly be more than a grain in any case.

54 **younker** young man, usually a fashion-
  able one

56 SIXTH WITCH The line is given to Firestone
  in *The Witch* (see note on 3.5.50), and to
  Hecate by Davenant, which demeans
  her.
  red-haired wench Red hair was often
  alluded to as poisonous, e.g. Chapman,
  Bussy D'Ambois, 3.2.18, 'the poison of a
  red-haired man'. Judas, the traitor, was
  represented as red-haired, so was Mary
  Magdalene, the prostitute, which may be
  suggested by the definite article here.

SECOND WITCH

By the pricking of my thumbs,
Something wicked this way comes—   60
Open locks, whoever knocks.

Enter Macbeth

MACBETH

How now, you secret, black, and midnight hags?
What is't you do?

ALL THE WITCHES   A deed without a name.

MACBETH

I conjure you, by that which you profess,
Howe'er you come to know it, answer me:
Though you untie the winds, and let them fight
Against the churches; though the yeasty waves
Confound and swallow navigation up;
Though bladed corn be lodged, and trees blown down,
Though castles topple on their warders' heads;   70
Though palaces and pyramids do slope
Their heads to their foundations; though the treasure
Of nature's germen tumble all together,

73 germen] F (Germaine); germains POPE; Germins THEOBALD *conj.*; germens GLOBE
all together] F (altogether)

59 **pricking** tingling
63, 81, 104, 118, 125 ALL THE WITCHES
Probably only the first three witches
speak.
63 **A deed without a name** 'Naming' conferred identity, and sanctity, as in bap-
tism; to be without one was to be alienated from the Name (of God), and bas-
tardy was associated with the Devil. But
the phrase is more resonant than its
sources.
64 **that which you profess** i.e. foreknowledge
65 **Howe'er ... know it** i.e. by the black arts
67 **yeasty** turbid, frothy, as in fermentation
68 **navigation** shipping (this sense survived
in USA until 1850)
69 **lodged** beaten down
71 **slope** bend downwards
72-4 **the treasure ... sicken** A complex
climax to the sequence from l. 66, very
closely comparable to Lear's 3.2.1-9,
cursing the storm, except that where
Lear's was all in the imperative, Macbeth's is all conditional, governed by the
repeated 'though'. *Lear* 3.2.8-9 reads:
'Crack nature's moulds, all germens spill

at once, | That makes ingrateful man.'
'Germens' are germs, that is seeds, not of
disease, but of all creation, in Lear's line
'moulded' into all the forms of nature.
The general sense of Macbeth's lines is
the same, the chaos and destruction of all
the sources of the living universe, but the
syntax is less clear. 'treasure' means
treasury, the structure that contains the
germens, as well as the treasure that is
contained, and both 'tumble', though in
different senses: the building *altogether
collapses*, as the castles, palaces, and
pyramids have; nature's germens *all
tumble together* in a chaos of mutual
destruction. In effect, two separate sentences conclude simultaneously and lead
to the finality of 'Even till destruction
sicken'; Muir suggests that the simplest
gloss is that destruction sickens of surfeit,
but the obvious force is that destruction
destroys itself—the phrase is reminiscent
of Donne's 'Death thou shalt die' (*Divine
Meditations*, 10).
73 **germen** F's 'germaine' is simply an old
spelling; there is no need to read 'germens' as most editors do if we assume the

Even till destruction sicken: answer me
To what I ask you.

FIRST WITCH   Speak.

SECOND WITCH   Demand.

THIRD WITCH   We'll answer.

FIRST WITCH

Say, if th'hadst rather hear it from our mouths,
Or from our masters.

MACBETH   Call'em: let me see'em.

FIRST WITCH

Pour in sow's blood, that hath eaten
Her nine farrow; grease that's sweaten
From the murderer's gibbet, throw   80
Into the flame.

ALL THE WITCHES Come high or low:
Thy self and office deftly show.

>   *Thunder.*
>
>   *First Apparition, an armed head*

MACBETH

Tell me, thou unknown power—

FIRST WITCH   He knows thy thought:
Hear his speech, but say thou nought.

FIRST APPARITION

Macbeth, Macbeth, Macbeth: beware Macduff,
Beware the Thane of Fife. Dismiss me; enough.

>   *Descends*

81, 104, 118, 125 ALL THE WITCHES] F (All.)   86.1 *Descends*] F (*He Descends*)

possibility of a collective form, though
this is not recorded before 1759 (*OED* 3,
Bot., an ovary).

77 **our masters** Evidently the spirits (devils)
who impersonate the apparitions.

78–9 **that … farrow** Apparently sows that
eat their young were thought to be poisonous. See Holinshed, *History of Scotland*
(1577), itemizing the laws of Kenneth II:
'If a sowe eate hir pigges, let hyr be stoned
to death and buried.'

79 **farrow** litter of piglets
sweaten exuded (from the gibbet)

80 **gibbet** Not only used of gallows, but also
of the similar structure (upright post with
projecting arm) from which the bodies of

executed criminals were hung in chains.

82.2 **armed head** There have been various
interpretations of the symbolism of the
apparitions, including improbable assumptions that the armed head is Macbeth's, or Macduff's. They remain, as
they undoubtedly should, cryptic, but it
is true that the sequence 'armed head',
'bloody babe', 'crowned child holding a
tree' (presumably of fertility) suggests a
meaningful sequence through death to
rebirth—from which Macbeth will be
excluded. Since the first three apparitions
are all directed to 'descend', it seems most
likely that they come up on a trap (or
traps, but one would suffice), possibly
with smoke from below.

MACBETH
Whate'er thou art, for thy good caution, thanks;
Thou hast harped my fear aright. But one word
    more—

FIRST WITCH
He will not be commanded: here's another
More potent than the first.   90
    Thunder.
    Second Apparition, a bloody child

SECOND APPARITION Macbeth, Macbeth, Macbeth.

MACBETH Had I three ears, I'd hear thee.

SECOND APPARITION
Be bloody, bold, and resolute: laugh to scorn
The power of man; for none of woman born
Shall harm Macbeth.   *Descends*

MACBETH
Then live Macduff—what need I fear of thee?
But yet I'll make assurance double sure,
And take a bond of fate: thou shalt not live,
That I may tell pale-hearted fear it lies,
And sleep in spite of thunder.
    Thunder.
    Third Apparition, a child crowned, with a tree in his
    hand
                      What is this   100
That rises like the issue of a king,
And wears upon his baby-brow the round
And top of sovereignty?

ALL THE WITCHES Listen, but speak not to't.

THIRD APPARITION
Be lion-mettled, proud, and take no care
Who chafes, who frets, or where conspirers are:
Macbeth shall never vanquished be, until
Great Birnam Wood to high Dunsinan Hill

---

97 assurance ∧] POPE; assurance: F   108 Dunsinan] This edition; Dunsmane F; Dunsinane
ROWE

88 **harped** given voice to, guessed
92 **three ears** All the apparitions call him
    three times.
108 **Birnam Wood...Dunsinan Hill** F varies
    between 'Byrnam' and 'Byrnan' (Birnam,
    Birnane, Byrnane), and between 'Duns-

mane' (possibly a misreading of 'Dunsinane') and 'Dunsinane'; 'Birnam' seems
to be usual now; Ordnance Survey maps
have 'Dunsinnan', which places the
stress on the second syllable as it is in this
line, but later in the play it is on the third;

Shall come against him.   *Descends*

MACBETH   That will never be:

Who can impress the forest, bid the tree   110

Unfix his earthbound root? Sweet bodements, good—

Rebellious dead, rise never till the Wood

Of Birnam rise, and our high-placed Macbeth

Shall live the lease of nature, pay his breath

To time, and mortal custom. Yet my heart

Throbs to know one thing: tell me, if your art

Can tell so much—shall Banquo's issue ever

Reign in this kingdom?

ALL THE WITCHES   Seek to know no more.

MACBETH

I will be satisfied. Deny me this,

And an eternal curse fall on you—let me know.   120

                    ⌈*Cauldron descends*⌉

Why sinks that cauldron? (*Hautboys*) And what noise
          is this?

FIRST WITCH Show.

SECOND WITCH Show.

THIRD WITCH Show.

ALL THE WITCHES

Show his eyes, and grieve his heart;

Come like shadows, so depart.

          A show of eight kings, the last with a glass in his
          hand, and Banquo

109 *Descends*] F (*Descend.*)   112 dead] F; head THEOBALD   113 our high-placed] F (our high plac'd); on's high place OXFORD   120.1] *not in* F   121 *Hautboys*] F (*Hoboyes*) *at end of line* 126.1–2] HANMER (*subst.*); *A shew of eight Kings, and Banquo last, with a glasse in his hand* F

I have adopted 'Dunsinan' to allow the variation which was usual with all proper nouns. They are a few miles apart, about 20 miles north of Perth.

110 **impress** enlist for military service
111 **bodements** predictions
112 **Rebellious dead** dead that will not stay buried; it could refer to all Macbeth's victims, but Banquo is the obvious one. Theobald's emendation to 'head', still often followed, is as unconvincing as it is unnecessary.
113 **our high-placed Macbeth** Macbeth retains the royal 'we' as long as he is 'high-placed'; compare 'my' and 'me' in ll. 115–16. It has been thought odd that

he should speak of himself in this way, but he is contemplating himself objectively as 'high-placed' both here and with 'his breath' in the next line. Oxford's emendation does not seem to me an improvement.

114 **lease of nature** natural lease of life. See proverb 'no man has lease of his life', Dent M327.
114–15 **pay his breath | To time** breathe till his time is due
115 **mortal custom** customary death
117 **ever** always
121 **noise** often used of agreeable sounds, especially from a group of musicians
126.1–2 F's direction is no doubt a conflation of separate marginal notes, confused by

MACBETH

Thou art too like the spirit of Banquo: down—
Thy crown does sear mine eye-balls. And thy hair,
Thou other gold-bound-brow, is like the first;
A third, is like the former. Filthy hags,   130
Why do you show me this?—A fourth? Start eyes!
What, will the line stretch out to th' crack of doom?
Another yet? A seventh? I'll see no more—
And yet the eighth appears, who bears a glass
Which shows me many more; and some I see
That two-fold balls, and treble sceptres carry.
Horrible sight—now I see 'tis true,
For the blood-baltered Banquo smiles upon me,
And points at them for his. What, is this so?
⌐*Exeunt kings and Banquo*⌐

⌐HECATE⌐

Ay sir, all this is so. But why   140
Stands Macbeth thus amazèdly?
Come sisters, cheer we up his sprites,
And show the best of our delights.
I'll charm the air to give a sound,
While you perform your antic round:
That this great King may kindly say,

134 eighth] F (eight)   138 blood-baltered] F (Blood-bolter'd)   139.1] GLOBE (*subst., after* his); *not in* F   140 HECATE] OXFORD (*conj.* Cambridge); I F

copyist or compositor; it has to be emended to fit ll. 134 and 138.

126.1 *eight Kings* A correct count of the Stuart kings of Scotland, but omitting Mary Queen of Scots, who was James's mother, beheaded by Elizabeth, his predecessor on the English throne. See Introduction, pp. 73–4. There is no need for eight actors if they move round backstage and re-enter with different emblems (depending on the structure of the theatre).

136 **two-fold balls, and treble sceptres** Referring to James's unifying of the kingdoms of Scotland and England: Scottish kings were invested with one sceptre and one orb, English kings with two sceptres and one orb.

138 **blood-baltered** blood-matted hair; or simply 'clotted'—*OED* gives 'bolter' as a spelling of 'balter' in this sense

139.1 *Exeunt ... Banquo* F gives no exit: most, if not all, of the kings have no doubt disappeared already.

140 HECATE F gives this to the First Witch, but it is generally agreed to be in Hecate's distinctive tone, and it introduces another dance, which seems to be her prerogative. Hecate, once she is there, becomes in a sense First Witch, and there may be confusion in the prefixes (see E. B. Lyle, 'The Speech-Heading "I" in Act IV Scene I of the Folio Text of *Macbeth*', *The Library*, 25 (1970), 150–1). The speech probably displaced an original one for the First Witch (see note on l. 38.1)

145 **antic round** grotesque dance

146–7 The conventional dedication of a masque: no doubt it could be dedicated to James at a performance in his presence, but it is here dedicating an anti-masque to Macbeth.

Our duties did his welcome pay.
> *Music. The Witches dance, and vanish ⌈with*
> *Hecate⌉*

MACBETH
Where are they? Gone? Let this pernicious hour
Stand aye accursèd in the calendar.
Come in, without there.
> *Enter Lennox*

LENNOX   What's your grace's will?   150

MACBETH
Saw you the Weïrd Sisters?

LENNOX   No, my lord.

MACBETH
Came they not by you?

LENNOX   No indeed, my lord.

MACBETH
Infected be the air whereon they ride,
And damned all those that trust them. I did hear
The galloping of horse. Who was't came by?

LENNOX
'Tis two or three, my lord, that bring you word:
Macduff is fled to England.

MACBETH   Fled to England?

LENNOX Ay, my good lord.

MACBETH (*aside*)
Time, thou anticipat'st my dread exploits;
The flighty purpose never is o'ertook   160
Unless the deed go with it. From this moment,
The very firstlings of my heart shall be
The firstlings of my hand. And even now

147.1–2 *with Hecate*] *not in* F   151 Weïrd] F (Weyard)

147.1 *The Witches dance* See Appendix B.
  with Hecate F gives no exit for her; many
  editors, and some stage-directors, take
  her off after l. 43, but it seems far more
  likely that she presides over the whole
  show and vanishes with the dance.
150 **without there** i.e. you who are outside
  there
155 **horse** horses (a collective plural still in
  use derived from OE)
157 See Introduction, p. 52.

158 The short line suggests a pause before
  Macbeth's soliloquy.
159 **anticipat'st** forestallest
160–1 **The flighty ... go with it** 'the flying
  (swift) purpose will never be caught up
  with unless the deed follows as quickly as
  it'; i.e. an intention must be carried out
  instantly or it will be too late
162 **firstlings** first concepts (literally 'first
  offspring')

To crown my thoughts with acts—be it thought and
   done:
The castle of Macduff I will surprise,
Seize upon Fife; give to th' edge o'th' sword
His wife, his babes, and all unfortunate souls
That trace him in his line. No boasting like a fool,
This deed I'll do, before this purpose cool;
But no more sights.—Where are these gentlemen?   170
Come bring me where they are.   *Exeunt*

4.2   *Enter Macduff's Wife, her Son, and Ross*
LADY MACDUFF
  What had he done, to make him fly the land?
ROSS
  You must have patience, madam.
LADY MACDUFF   He had none—
  His flight was madness: when our actions do not,
  Our fears do make us traitors.
ROSS   You know not
  Whether it was his wisdom, or his fear.
LADY MACDUFF
  Wisdom? To leave his wife, to leave his babes,
  His mansion, and his titles, in a place
  From whence himself does fly? He loves us not,
  He wants the natural touch. For the poor wren,
  The most diminutive of birds, will fight,   10
  Her young ones in her nest, against the owl.
  All is the fear, and nothing is the love;
  As little is the wisdom, where the flight
  So runs against all reason.
ROSS   My dearest coz,
  I pray you school yourself. But for your husband,
  He is noble, wise, judicious, and best knows

---

164 **be it thought and done** Variation on 'no
    sooner said than done', Dent S117.
168 **trace him in his line** follow him in his
    lineage, i.e. his relatives and, possibly,
    dependants
170 **But no more sights** The passionate
    appeal for an end to visions is heard:
    his sight from here on is as literal
    as the moving of Birnam Wood; see

Introduction, p. 5.
4.2.4 **Our fears do make us traitors** i.e. by
    fleeing and thus seeming to admit
    treachery—as Malcolm, Donalbain,
    and Fleance have all done
  7 **titles** entitlement, both to his thaneship
    and to the estates that go with it
 15 **school** control

The fits o'th' season. I dare not speak much further,
But cruel are the times, when we are traitors
And do not know ourselves; when we hold rumour
From what we fear, yet know not what we fear, 20
But float upon a wild and violent sea
Each way and move.—I take my leave of you:
Shall not be long but I'll be here again—
Things at the worst will cease, or else climb upward
To what they were before. (*To Son*) My pretty cousin,
Blessing upon you.

LADY MACDUFF
Fathered he is, and yet he's fatherless.

ROSS
I am so much a fool, should I stay longer
It would be my disgrace, and your discomfort.
I take my leave at once.   *Exit* 30

LADY MACDUFF Sirrah, your father's dead, and what will
you do now? How will you live?

SON As birds do, mother.

LADY MACDUFF What, with worms and flies?

SON With what I get, I mean, and so do they.

LADY MACDUFF Poor bird, thou'dst never fear the net, nor

4.2.22 way and move.] This edition; way, and moue. F; way and wave. THEOBALD; way and
move—JOHNSON; way and none. WILSON (*conj.* Cambridge), OXFORD; *numerous other conjectures*
30 *Exit*] F (*Exit Rosse*)

17 **fits o'th' season** violent disorders of the
time. See *Coriolanus* 3.2.33, 'The violent
fit o'th' time'; 'fits' might also imply
'appropriate', but *OED* has that sense
only for the verb before 1688.

19 **do not know ourselves** i.e. are bewildered, when loyalty is so inevitably
divided that the most faithful to the
country may be traitors to the King.

22 **way and move** My punctuation assumes
that 'move' is a noun, in the rare sense of
'motion' (similar to such phrases as 'get a
move on'), i.e. (float upon) every way and
movement (of the sea). If Johnson was
right, as many editors think, 'way, and
move—' would imply that Ross breaks off
abruptly, to stop himself saying any
more; this is very possible, and similar in
form to Macbeth's 'And falls on th'
other—' in 1.7.28, which F also closed
with a full stop. Empson suggests that
'move' operates as a distanced and calm

version of the agitation of the last four
lines (*Seven Types*, p. 101). None of these
need be excluded, and in any case,
'move' finally suggests departure (*OED*
3—as in 'make a move').

24 **Things ... cease** See sententious 'when
things are at their worst they will mend',
Dent T216.

28–9 **a fool ... my disgrace** To weep is not a
man's role, and Ross would weep if he
stayed.

31–53 See Appendix A.

33 **As birds do** Editors refer to Matthew 6:
26, 'Behold the fowls of the air: for they
sow not, neither do they reap ... yet your
heavenly father feeds them.' The idea is
similar, but there is no verbal echo.

36, 38 **Poor bird ... Poor birds** Playing
on 'Poor': 1. unfortunate; 2. poverty-
stricken, starving (traps are set for fat
birds).

lime, the pitfall, nor the gin.

SON Why should I, mother? Poor birds they are not set
for: my father is not dead for all your saying.

LADY MACDUFF Yes, he is dead: how wilt thou do for a   40
father?

SON Nay how will you do for a husband?

LADY MACDUFF Why, I can buy me twenty at any market.

SON Then you'll buy 'em to sell again.

LADY MACDUFF Thou speak'st with all thy wit, and yet
i'faith with wit enough for thee.

SON Was my father a traitor, mother?

LADY MACDUFF Ay, that he was.

SON What is a traitor?

LADY MACDUFF Why, one that swears, and lies.   50

SON And be all traitors, that do so?

LADY MACDUFF Everyone that does so is a traitor, and must
be hanged.

SON And must they all be hanged, that swear and lie?

LADY MACDUFF Every one.

SON Who must hang them?

LADY MACDUFF Why, the honest men.

SON Then the liars and swearers are fools: for there are
liars and swearers enough to beat the honest men, and
hang up them.   60

LADY MACDUFF Now God help thee, poor monkey! But
how wilt thou do for a father?

SON If he were dead, you'd weep for him; if you would
not, it were a good sign that I should quickly have a
new father.

LADY MACDUFF Poor prattler, how thou talk'st!

Enter a Messenger

---

59 enough] F (enow)

---

37 **lime** bird-lime, glue painted on trees to
catch birds
gin trap
42–4 Dent B787 refers this to the proverb 'to
be bought and sold', meaning to deceive
or be deceived.
45–6 **Thou speak'st ... for thee** You speak
with as much wisdom as you possess, and
yet in truth you're shrewd enough; playing on senses of 'wit', including 'witty'.
50 **swears, and lies** swears, and proves his
oaths false
59 **enough** F printed 'enough' in l. 46,
'enow' here; it may mark the boy's
colloquial speech.

MESSENGER

Bless you fair dame: I am not to you known,
Though in your state of honour I am perfect;
I doubt some danger does approach you nearly.
If you will take a homely man's advice,   70
Be not found here: hence with your little ones;
To fright you thus methinks I am too savage:
To do worse to you were fell cruelty,
Which is too nigh your person. Heaven preserve you,
I dare abide no longer.   *Exit*

LADY MACDUFF   Whither should I fly?
I have done no harm. But I remember now
I am in this earthly world, where to do harm
Is often laudable, to do good sometime
Accounted dangerous folly. Why then, alas,
Do I put up that womanly defence,   80
To say I have done no harm?
        Enter Murderers
What are these faces?

MURDERER   Where is your husband?

LADY MACDUFF

I hope in no place so unsanctified
Where such as thou mayst find him.

MURDERER   He's a traitor.

SON

Thou li'st, thou shag-eared villain.

71–2 little ones; | . . . thus methinks] F2 (*subst.*); little ones | . . . thus. Me thinkes F1   75 *Exit*]
F (*Exit Messenger*)   81.1 *Enter Murderers*] F, *after* faces?   85 shag-eared] F (shagge-ear'd);
shag-haired SINGER (*conj.* Steevens)

---

68 **in your state . . . perfect** I am fully aware
of your honourable status and reputation
69 **doubt** dread
72–4 **To fright . . . person** i.e. 'it is savage
enough to frighten you, actually to harm
you physically would be downright besti-
ality—and that is already too close to
you'. Other explanations of this elliptical
passage do not help much—its obscurity
derives from the Messenger's reluctance
to be more explicit.
81–2 The short lines indicate that Lady
Macduff breaks off, probably in response

to a noise off-stage, which may cause her
to see the Murderers as they enter;
alternatively, they may enter earlier,
behind her. See Appendix A (80–2).
85 **shag-eared** There seems no point in
emending to 'shag-haired', even it if is
true that that was a cliché for villains and
could be spelt 'shag-heared'; 'shag-
eared' is apt for curs, and convicted
villains frequently had their ears slit.
Doubtless both words can be heard at
once.

MURDERER   What, you egg?
Young fry of treachery! ⌈*Kills him*⌉
SON   He has killed me, mother,
Run away I pray you.
          Exit ⌈*Lady Macduff*⌉ *crying 'Murder',* ⌈*pursued by*
                             Murderers with her Son⌉

4.3   *Enter Malcolm and Macduff*
MALCOLM
Let us seek out some desolate shade, and there
Weep our sad bosoms empty.
MACDUFF   Let us rather
Hold fast the mortal sword—and like good men,
Bestride our downfall birthdom; each new morn
New widows howl, new orphans cry, new sorrows
Strike Heaven on the face, that it resounds
As if it felt with Scotland, and yelled out
Like syllable of dolour.
MALCOLM   What I believe, I'll wail;
What know, believe; and what I can redress,
As I shall find the time to friend—I will.   10
What you have spoke, it may be so perchance.
This tyrant, whose sole name blisters our tongues,
Was once thought honest: you have loved him well,

86 *Kills him*] not in F   87.1–2] This edition; *Exit crying Murther.* F
    4.3.4 downfall] F; downfall'n WARBURTON

85–6 See proverb 'an ill bird lays an ill egg', Dent B376.
85, 86 **egg, fry** infant (Costard calls Mote 'pigeon-egg' in *LLL* 5.1.70; 'fry' was used for human offspring, but also (as now) of fish-spawn and newly hatched fish.
86 *Kills him* presumably by stabbing or strangling
87.2 **with her Son** It is not clear how the Son is got off stage: presumably he either staggers after the Murderers, or is dragged by them.
4.3.3 **mortal** deadly
4 **Bestride our downfall birthdom** The metaphor is clearly of bestriding the body of a slaughtered friend, challenging his enemy. 'birthdom' is equivalent to 'birthright'; 'downfall' may be a past participle, i.e. 'downfall'n', or the noun doubling as an adjective, i.e. 'bestride our own downfall, *and* bestride our destroyed birthright'.
6 **that** so that
   it i.e. heaven in the sense of the dome of the sky (as well as the 'face of heaven' that is struck and yells)
8 **Like syllable** similar cry; 'syllable' referred to a single sound, e.g. a howl, etc., before it came to be used only of a part of a word
   dolour lamentation
8 ff. Malcolm's language has the formality of courtesy complicated by suspicion (see 3.6).
10 **time to friend** time friendly, i.e. favourable opportunity
12 See proverb 'report has a blister on her tongue', Dent R84
   sole name name alone

He hath not touched you yet. I am young, but
    something
You may discern of him through me, and wisdom
To offer up a weak, poor, innocent lamb
T'appease an angry God.

MACDUFF
I am not treacherous.

MALCOLM   But Macbeth is.
A good and virtuous nature may recoil
In an imperial charge. But I shall crave your pardon: 20
That which you are, my thoughts cannot transpose;
Angels are bright still, though the brightest fell.
Though all things foul would wear the brows of grace,
Yet grace must still look so.

MACDUFF   I have lost my hopes.

MALCOLM
Perchance even there where I did find my doubts.
Why in that rawness left you wife, and child—
Those precious motives, those strong knots of love—
Without leave-taking? I pray you,
Let not my jealousies be your dishonours,
But mine own safeties; you may be rightly just,   30
Whatever I shall think.

MACDUFF   Bleed, bleed, poor country—
Great tyranny, lay thou thy basis sure,
For goodness dare not check thee: wear thou thy
    wrongs,

---

15 discern] F; deserve THEOBALD

14–15 **something ... me** you may discern
elements of him in me (see ll. 46 ff.);
Theobald's 'deserve' may seem simpler,
but it makes poor sense of 'I am young,
but ...'.

15 **and wisdom** i.e. you may discern the
wisdom

16 **weak ... lamb** The irony is obvious; 'as
innocent as a lamb' was already
proverbial (Dent L234.1).

19 **recoil** retreat (from its virtue)

20 **imperial** 1. royal; 2. imperious
charge 1. command; 2. attack

21 **transpose** change, corrupt

23–4 **Though ... so** the fact that hideous

things may mask themselves as gracious
does not alter the necessity for grace to
look gracious

24 **I have lost my hopes** i.e. because trust is
impossible

27 **motives** Often used of the people who
initiate action, especially through love.

28 The missing foot in the line is supplied by
the natural pause after 'leave-taking?',
before a change of tone in 'I pray you'—
especially if Macduff reacts violently.

29–30 **Let not ... safeties** i.e. Don't take my
suspicions to be necessarily insults to
you, so much as precautions for my own
safety

The title is affeered. Fare thee well lord,
I would not be the villain that thou think'st
For the whole space that's in the tyrant's grasp,
And the rich East to boot.

MALCOLM   Be not offended:
I speak not as in absolute fear of you—
I think our country sinks beneath the yoke,
It weeps, it bleeds, and each new day a gash   40
Is added to her wounds. I think withal
There would be hands uplifted in my right;
And here from gracious England have I offer
Of goodly thousands. But for all this,
When I shall tread upon the tyrant's head,
Or wear it on my sword—yet my poor country
Shall have more vices than it had before,
More suffer, and more sundry ways than ever,
By him that shall succeed.

MACDUFF   What should he be?

MALCOLM
It is myself I mean—in whom I know   50
All the particulars of vice so grafted,
That when they shall be opened, black Macbeth
Will seem as pure as snow, and the poor state
Esteem him as a lamb, being compared
With my confineless harms.

MACDUFF   Not in the legions
Of horrid Hell can come a devil more damned
In evils, to top Macbeth.

MALCOLM   I grant him bloody,
Luxurious, avaricious, false, deceitful,
Sudden, malicious, smacking of every sin

---

34 affeered] F (affear'd)

34 **title** 1. tyranny; 2. claim to the throne
affeered 1. confirmed (a legal term); 2.
'affeared' (F's spelling), frightened (of
pressing his claim)
43 **England** commonly used for the King of
England
48 **more sundry** (suffer in) more different
51 **grafted** as in the culture of trees, etc.
52 **opened** 1. like blossoms from the graft-
ing; 2. revealed
52–3 **black ... snow** Dent refers to two

proverbs: 'to make black white' (B440)
and 'as pure as snow' (S591).
58 **Luxurious** lecherous (as most commonly
then); Malcolm's catalogue of sins is
conventional and not, as has been often
suggested, a characterization of Macbeth.
59 **Sudden** hasty (but with a stronger sense
of violence); Gary Taylor suggests that
Compositor B misread 'sullen' as 'sudden' (which he usually prints 'sodaine',
as here); 'sullen' had the sense 'malig-

That has a name. But there's no bottom, none,   60
In my voluptuousness: your wives, your daughters,
Your matrons, and your maids, could not fill up
The cistern of my lust; and my desire
All continent impediments would o'erbear
That did oppose my will. Better Macbeth,
Than such an one to reign.

MACDUFF   Boundless intemperance
In nature is a tyranny: it hath been
Th'untimely emptying of the happy throne,
And fall of many kings. But fear not yet
To take upon you what is yours; you may   70
Convey your pleasures in a spacious plenty,
And yet seem cold. The time you may so hoodwink—
We have willing dames enough; there cannot be
That vulture in you, to devour so many
As will to greatness dedicate themselves,
Finding it so inclined.

MALCOLM   With this, there grows
In my most ill-composed affection, such
A stanchless avarice, that were I King,
I should cut off the nobles for their lands,
Desire his jewels, and this other's house,   80
And my more-having would be as a sauce
To make me hunger more, that I should forge
Quarrels unjust against the good and loyal,
Destroying them for wealth.

---

nant', but does not seem to me an improvement.

63 **cistern** Used of a very large tank, a pond, or any natural reservoir; for the metaphor of insatiable lust, see also *Othello* 4.2.6: 'keep it as a cistern for foul toads | To knot or gender in'.

64 **continent impediments** Literally 'restraining obstructions', but 'continent' meant 'chaste', so the phrase is not simply tautologous.

65 **will** appetitive will, i.e. lust

66–7 **Boundless … tyranny** It is tyranny if the appetites, or 'will', take control of the brain which should be 'king' of the body. The analogy helps to define 'tyranny' for which alone, many theorists held, a king

could be deposed.

71 **Convey** carry on secretly; 'convey' was often used of theft
in a spacious plenty The construction is analogous to 'in a large way'.

75 **to greatness dedicate** devote, or surrender, to a powerful person; Macduff insinuates that plenty of people were ready to prostitute themselves to a king

77 **affection** disposition; commonly used of disorderly affections, lust etc.

78 **stanchless** without a stopper

79 **cut off** bring to an untimely end, i.e. kill

80 **his** i.e. 'this one's'

81–2 **my more-having … more** See proverb 'the more a man has the more he desires', Dent M1144.

MACDUFF　　　　　　　　　This avarice
　　Sticks deeper, grows with more pernicious root
　　Than summer-seeming lust; and it hath been
　　The sword of our slain kings; yet do not fear,
　　Scotland hath foisons to fill up your will
　　Of your mere own. All these are portable,
　　With other graces weighed.　　　　　　　　　　90

MALCOLM
　　But I have none. The king-becoming graces,
　　As justice, verity, temp'rance, stableness,
　　Bounty, perseverance, mercy, lowliness,
　　Devotion, patience, courage, fortitude,
　　I have no relish of them, but abound
　　In the division of each several crime,
　　Acting it many ways. Nay, had I power, I should
　　Pour the sweet milk of concord into Hell,
　　Uproar the universal peace, confound
　　All unity on earth.

MACDUFF　　　　　　　　O Scotland, Scotland.　　　　100

MALCOLM
　　If such a one be fit to govern, speak:
　　I am as I have spoken.

MACDUFF　　　　　　　　Fit to govern?
　　No, not to live. O nation miserable!
　　With an untitled tyrant, bloody-sceptered,
　　When shalt thou see thy wholesome days again?
　　Since that the truest issue of thy throne

---

86 **summer-seeming** 1. summer-like, i.e. lust is a relatively attractive vice, avarice is not; 2. beseeming summer, i.e. a young man's vice, supposed to wither with age, where avarice grows more powerful (from its pernicious root)

87 **The sword of our slain kings** Holinshed records this speech of Macduff's as 'avarice is the root of all mischiefe, and for that crime the most part of our kings have beene slain and brought to their final end'.

88 **foisons** resources ('foison' in the singular means 'plenty')

89 **your mere own** simply of your own (wealth as king)
portable bearable

93 **perseverance** Accented on the second syllable.
lowliness humility

96 **division** separate forms, possibly with a reference to the musical senses of dividing long notes into multiples of short ones, i.e. variations, or descants

99 **Uproar** throw into confusion

104 **untitled** This has been glossed 'having no legal right', but Macbeth's title was legal, since primogeniture is not stated to be an over-riding claim. Macduff may none the less be referring to that, or to the theory that tyranny disqualified a king and so deprived him of his original right, or he may be looking forward to rebellion deposing Macbeth but finding no clear successor if Malcolm is unfit.

By his own interdiction stands accused,
And does blaspheme his breed? Thy royal father
Was a most sainted king; the queen that bore thee,
Oft'ner upon her knees than on her feet,　　　　　　110
Died every day she lived. Fare thee well,
These evils thou repeat'st upon thyself
Hath banished me from Scotland. O my breast,
Thy hope ends here.

MALCOLM　　　　　　Macduff, this noble passion,
Child of integrity, hath from my soul
Wiped the black scruples, reconciled my thoughts
To thy good truth, and honour. Devilish Macbeth,
By many of these trains, hath sought to win me
Into his power; and modest wisdom plucks me
From over-credulous haste—but God above　　　　　120
Deal between thee and me; for even now
I put myself to thy direction, and
Unspeak mine own detraction: here abjure
The taints and blames I laid upon myself,
For strangers to my nature. I am yet
Unknown to woman, never was forsworn,
Scarcely have coveted what was mine own,
At no time broke my faith, would not betray
The Devil to his fellow, and delight
No less in truth than life. My first false speaking　　130
Was this upon myself. What I am truly
Is thine, and my poor country's to command;
Whither indeed, before thy here approach,

107 accused] F1 (accust); accurst F2　　133 thy here approach] F2; they heere approach F1;
thy here-approach POPE

107 **interdiction** legal restraint—in Scottish law, specifically for unsoundness of mind, etc.
　　accused The usual reading 'accursed' seems to me no improvement though it has been justified both as a possible misreading or as rendering a possible pronunciation of 'accursed' ('r' lost before consonants).
108 **blaspheme his breed** slanders his family, with obvious reference to the ecclesiastical sense of *blaspheme* as Duncan was 'most sainted'

111 **Died** i.e. mortified herself, or daily contemplated death as a warning in life; see 1 Corinthians 15: 31
113 **Hath ... Scotland** i.e. mean that I have no hope of returning to Scotland
118 **trains** tricks, stratagems
119 **modest wisdom** 1. moderate wisdom, i.e. the wisdom of moderation; 2. 'wisdom' is personified, 'modest' in the sense of cautious
　　plucks pulls back (*OED* 3)
133 **thy here approach** F's 'they' makes no sense and is presumably a slip; the phrase

Old Seyward with ten thousand warlike men
Already at a point, was setting forth—
Now we'll together, and the chance of goodness
Be like our warranted quarrel. Why are you silent?

MACDUFF
Such welcome and unwelcome things at once
'Tis hard to reconcile.

Enter a Doctor

MALCOLM   Well, more anon.
Comes the King forth, I pray you?   140

DOCTOR
Ay, sir—there are a crew of wretched souls
That stay his cure: their malady convinces
The great assay of art. But at his touch,
Such sanctity hath Heaven given his hand,
They presently amend.   ⌜*Exit*⌝

MALCOLM   I thank you, Doctor.

MACDUFF
What's the disease he means?

MALCOLM   'Tis called the Evil.
A most miraculous work in this good King,
Which often since my here remain in England
I have seen him do. How he solicits Heaven
Himself best knows; but strangely visited people,   150

---

145 *Exit*] F; *after Malcolm's response* CAPELL

is directly comparable to 'my here remain' in l. 148, i.e. 'thy coming here',
'my stay here'.

134 **Old Seyward** A son of the Earl of
Northumberland, and a strong supporter
of Edward the Confessor; there seems no
point in adopting the editorial spelling
'Siward' for a figure only familiar in this
context.

135 **at a point** OED glosses this phrase
'agreed, resolved', which could mean
here 'actually levied'; it is usually glossed
'fully prepared', which seems likely;
there is a similar phrase 'at point' used of
Lear's knights at *Lear* 1.4.325

136 **goodness** i.e. good fortune (deserved, as
the quarrel is warranted)

137 **warranted** legally justified

140 There is presumably a pause before this

short line as Malcolm turns to the Doctor.

140–59 The miraculous 'Touching' for the
'King's Evil'—scrofula—completes the
(off-stage) image of Edward the Confessor
as the Divine King, as opposed to the evil
image Malcolm conjured of himself as all-
too-human king. Macduff tries to cast
Duncan and Macbeth in equivalent roles,
but neither is so absolute. 'Touching' was
practised by all English monarchs until
the late eighteenth century; James had
mixed feelings about it, see Introduction,
p. 72.

142 **convinces** overcomes

143 **assay** attempt, best effort

145 *Exit* F's placing is not sacrosanct, but
the Doctor was not summoned, and may
leave without being dismissed.

148 **here remain** See note on l. 133.

150 **visited** afflicted

All swoll'n and ulcerous, pitiful to the eye,
The mere despair of surgery, he cures,
Hanging a golden stamp about their necks
Put on with holy prayers; and 'tis spoken,
To the succeeding royalty he leaves
The healing benediction. With this strange virtue,
He hath a heavenly gift of prophecy,
And sundry blessings hang about his throne
That speak him full of grace.

Enter Ross

MACDUFF   See who comes here.

MALCOLM

My countryman: but yet I know him not.   160

MACDUFF

My ever gentle cousin, welcome hither.

MALCOLM

I know him now. Good God betimes remove
The means that makes us strangers.

ROSS   Sir, amen.

MACDUFF

Stands Scotland where it did?

ROSS   Alas poor country,
Almost afraid to know itself. It cannot
Be called our mother, but our grave; where nothing
But who knows nothing, is once seen to smile;
Where sighs, and groans, and shrieks that rend the air
Are made, not marked; where violent sorrow seems

---

154 with] F1 *corrected sheets*; my with F1 *uncorrected*   160 not] F2; nor F1   168 rend] F
(rent)

---

152 **mere** absolute
   surgery treatment, not necessarily with
   the knife
153 **stamp** coin or medal (made with a
   stamp)
154 **Put on with** The only case of verbal
   correction in the printing of *Macbeth* in F,
   and presumably just an obvious guess
   about the misprint.
   spoken said
155 **leaves** bequeathes
156 **virtue** quality of divine power

160 **My countryman** i.e. a Scot—presumably some distinctive elements of costume were used in the theatre to make
   this plausible.
162 **betimes** speedily
167 **But who knows nothing** except those
   who know nothing, i.e. children or
   idiots
168–9 **shrieks ... Are made, not marked** i.e.
   the most horrid shrieks happen but are
   not noticed

A modern ecstasy—the deadman's knell   170
Is there scarce asked for who, and good men's lives
Expire before the flowers in their caps,
Dying or ere they sicken.
MACDUFF   O relation,
Too nice and yet too true.
MALCOLM   What's the newest grief?
ROSS
That of an hour's age doth hiss the speaker,
Each minute teems a new one.
MACDUFF   How does my wife?
ROSS
Why well.
MACDUFF   And all my children?
ROSS   Well too.
MACDUFF
The tyrant has not battered at their peace?
ROSS
No, they were well at peace when I did leave 'em.
MACDUFF
Be not a niggard of your speech: how goes't?   180
ROSS
When I came hither to transport the tidings
Which I have heavily borne, there ran a rumour
Of many worthy fellows that were out,
Which was to my belief witnessed the rather
For that I saw the tyrant's power afoot.
Now is the time of help: your eye in Scotland
Would create soldiers, make our women fight,
To doff their dire distresses.
MALCOLM   Be't their comfort

170 **modern** commonplace; a short-lived usage which *OED* dates 1591–1610, particularly in Shakespeare: presumably it came from 'nowadays' via 'every day'; see *Romeo* 3.2.120.
170–1 **the deadman's knell ... for who** Alluding to the customary question 'for whom tolls the bell?' asked of the tolling after a death.
173 **or ere** before
174 **nice** 1. strange; 2. over-refined (i.e. rhetorically)

175 **doth hiss the speaker** causes the speaker to be hissed (because it's stale news)
176 **teems** gives birth to
179 **well at peace** Equivocating with 'at peace' applied to the deed. See proverb 'he is well since he is in heaven', Dent H347.
183 **out** i.e. in active rebellion (similar to the modern usage about industrial strikes)
185 **power** army
186 **your eye** i.e. the sight of you (Malcolm)
188 **doff** put aside

We are coming thither—gracious England hath
Lent us good Seyward, and ten thousand men,   190
An older and a better soldier none
That Christendom gives out.

ROSS   Would I could answer
This comfort with the like. But I have words
That would be howled out in the desert air,
Where hearing should not latch them.

MACDUFF   What concern they,
The general cause, or is it a fee-grief
Due to some single breast?

ROSS   No mind that's honest
But in it shares some woe, though the main part
Pertains to you alone.

MACDUFF   If it be mine
Keep it not from me, quickly let me have it.   200

ROSS
Let not your ears despise my tongue forever,
Which shall possess them with the heaviest sound
That ever yet they heard.

MACDUFF   H'm—I guess at it.

ROSS
Your castle is surprised; your wife, and babes,
Savagely slaughtered—to relate the manner
Were on the quarry of these murdered deer
To add the death of you.

MALCOLM   Merciful Heaven—
What man, ne'er pull your hat upon your brows:
Give sorrow words; the grief that does not speak
Whispers the o'erfraught heart, and bids it break.   210

MACDUFF
My children too?

192 **gives out** reports
195 **latch** take hold of
196 **fee-grief** The phrase seems to have
been coined here: 'fee' was an hereditary estate (often large), hence this grief
comes to only a single owner.
202 **possess them with** 1. put them in
possession of (*OED* II.7); 2. cause them to
be possessed by (*OED* II.9)
206 **quarry** the object of a hunt; hence, a
heap of deer killed in a hunt (*OED*, sb¹, 2)

deer 1. deer; 2. dear ones
209–10 **Give sorrow . . . heart** A version of a
favourite tag, possibly proverbial, 'grief
pent up will break the heart' (Dent
G449); though usually traced to Seneca,
Hippolytus, 607, 'Curae leves loquuntur,
ingentes stupent' (Light cares speak out,
enormous ones stun).
210 **Whispers the o'erfraught heart** i.e.
speaks secretly only to the overburdened
heart

ROSS   Wife, children, servants, all
That could be found.
MACDUFF   And I must be from thence!
My wife killed too?
ROSS   I have said.
MALCOLM   Be comforted.
Let's make us med'cines of our great revenge,
To cure this deadly grief.
MACDUFF
He has no children. All my pretty ones?
Did you say all? O hell-kite! All?
What, all my pretty chickens and their dam
At one fell swoop?
MALCOLM   Dispute it like a man.
MACDUFF I shall do so: 220
But I must also feel it as a man;
I cannot but remember such things were
That were most precious to me; did Heaven look on,
And would not take their part? Sinful Macduff,
They were all struck for thee—naught that I am,
Not for their own demerits but for mine
Fell slaughter on their souls: Heaven rest them now.
MALCOLM
Be this the whetstone of your sword, let grief
Convert to anger: blunt not the heart, enrage it.
MACDUFF
O I could play the woman with mine eyes,   230

214–15 Muir comments that one passion was thought to drive out another, but here the suggestion is rather that revenge is an outlet for silent grief (see ll. 209–10), as in the proverb 'to lament the dead avails not and revenge vents hatred' (Dent D125); the idea is the burden of Hieronimo's decision to revenge in Kyd's *Spanish Tragedy*, 3.13.1–20; Malcolm tried to use Macduff's grief for his own purposes.

215–20 See Appendix A. Most editors rearrange to reduce the number of short lines, but the F arrangement gives all the pauses to Macduff, which is appropriate.

216 **He has no children** 1. Malcolm would not offer such a simplistic cure if he had children of his own; 2. Revenge on Macbeth's children is impossible because he has none; 3. If Macbeth had children, he would not have slaughtered others. The first sense seems to me an inevitable snub to Malcolm's glib haste. See proverb 'he that has no children knows not what love is', Dent C341.

218 **dam** mother

219 **one fell swoop** 'swoop' is the pounce of a bird of prey: the phrase derives from hawking; it is not found as a proverb before *Macbeth*

229 **blunt ... enrage** i.e. let grief blunt ... enrage

230 **play the woman** proverbial phrase, Dent W637.2

And braggart with my tongue. But gentle heavens,
Cut short all intermission: front to front
Bring thou this fiend of Scotland and myself,
Within my sword's length set him—if he scape,
Heaven forgive him too.
MALCOLM   This time goes manly—
Come, go we to the King, our power is ready,
Our lack is nothing but our leave. Macbeth
Is ripe for shaking, and the powers above
Put on their instruments; receive what cheer you may,
The night is long that never finds the day.   *Exeunt*   240

5.1   *Enter a Doctor of Physic, and a Waiting Gentlewoman*
DOCTOR I have two nights watched with you, but can
perceive no truth in your report. When was it she last
walked?

GENTLEWOMAN Since his majesty went into the field, I
have seen her rise from her bed, throw her night-gown
upon her, unlock her closet, take forth paper, fold it,
write upon't, read it, afterwards seal it, and again
return to bed; yet all this while in a most fast sleep.

DOCTOR A great perturbation in nature, to receive at once
the benefit of sleep, and do the effects of watching. In   10

235 time] F; tune ROWE 1714

231 **braggart** The primary sense of 'brag' is
'bray', the loud noise of a trumpet, which
seems to be the primary sense of 'braggart' here, though according to *OED*
braggart is derived from French *bragard*
and not from brag; the sense 'vain
boaster' is also present, implying empty
windiness.

232 **intermission** interval, 1. of time, 2. of
space—a common Latin sense which was
rarely used in English in the seventeenth
century, but seems here to govern Macduff's next words

235 **This time goes manly** i.e. 1. now ('this
time') we are ready for action; 2. this
march rhythm ('time') is manly (see
ll. 230-1)
   time Most editors since Rowe have
emended to 'tune' which makes easier
sense but is not necessarily correct; the
musical metaphor runs from 'braggart'

through 'intermission', 'time' (tune),
into 'instruments' in l. 239. Tilley referred to proverb 'times change and we
with them', Dent T343.

236 **power** army (see l. 185)

237 **Our lack ... leave** all we need to do is
formal leave-taking

238 **powers above** celestial influences,
angels

239 **Put on their instruments** 1. blow their
trumpets; 2. put on their weapons; 3. set
us, their instruments, on (to achieve their
divine purpose)

5.1.4 **field** battlefield

5 **night-gown** dressing-gown

6 **closet** cabinet (*OED* 3, citing *Lear* 3.3.11,
where it is also used for letters)
   paper What she writes and seals is left
uncertain—a note to Macbeth, a confession, a will?

10 **watching** being awake

this slumbery agitation, besides her walking and other
actual performances, what, at any time, have you heard
her say?

GENTLEWOMAN That, sir, which I will not report after her.

DOCTOR You may to me, and 'tis most meet you should.

GENTLEWOMAN Neither to you, nor anyone, having no
witness to confirm my speech.

Enter Lady Macbeth as Queen, with a taper

Lo you, here she comes—this is her very guise, and
upon my life fast asleep; observe her, stand close.

DOCTOR How came she by that light? 20

GENTLEWOMAN Why, it stood by her; she has light by her
continually, 'tis her command.

DOCTOR You see her eyes are open.

GENTLEWOMAN Ay but their sense are shut.

DOCTOR What is it she does now? Look how she rubs her
hands.

GENTLEWOMAN It is an accustomed action with her, to
seem thus washing her hands: I have known her
continue in this a quarter of an hour.

LADY MACBETH Yet here's a spot. 30

DOCTOR Hark, she speaks, I will set down what comes from
her, to satisfy my remembrance the more strongly.

LADY MACBETH Out damned spot—out I say. One—two—
why then 'tis time to do't—Hell is murky. Fie, my lord,
fie, a soldier, and afeard? What need we fear who knows
it, when none can call our power to account? Yet who

5.1.17.1 *Lady Macbeth as Queen*] F (*Lady*)   24 sense are] F; sense is ROWE   35 fear ∧ who]
THEOBALD; feare? who F

11 **agitation** 1. perturbation of mind; 2.
    perturbed activity
12 **actual** active (*OED* 1)
14 **after** In the sense of 'in pursuit of', i.e. as
    an accusation.
18 **guise** customary manner
19 **close** hidden
24 **their sense are** 'sense' may, as Oxford
    claims, have been used as a plural; *OED*
    does not have this, but does give the
    singular collective, as in *Othello* 4.3.93
    (where sight, smell, and taste are spe-
    cified and the other senses implied). Here
    the plural 'are' may be influenced by

there being two eyes, as in Sonnet 112 it
probably was by two ears: 'my adder's
sense | To critic and to flatterer stoppèd
are'; but in both contexts it may also
have been suggested by other plural
forms, in this case 'their'.
32 **satisfy my remembrance** set free from
    uncertainty (*OED* II.4) my remembering
33-4 **One—two—why then 'tis time** Pre-
    sumably a clock striking (in her imagina-
    tion); no clock is mentioned in 2.1,
    where there is a bell (ll. 33 and 63) but no
    suggestion that there is any significance
    in how many times it rings.

would have thought the old man to have had so much blood in him.

DOCTOR Do you mark that?

LADY MACBETH The Thane of Fife, had a wife—where is she now? What, will these hands ne'er be clean? No more o'that my lord, no more o'that—you mar all with this starting.   40

DOCTOR Go to, go to: you have known what you should not.

GENTLEWOMAN She has spoke what she should not, I am sure of that; Heaven knows what she has known.

LADY MACBETH Here's the smell of the blood still—all the perfumes of Arabia will not sweeten this little hand. O, O, O.   50

DOCTOR What a sigh is there! The heart is sorely charged.

GENTLEWOMAN I would not have such a heart in my bosom, for the dignity of the whole body.

DOCTOR Well, well, well.

GENTLEWOMAN Pray God it be, sir.

DOCTOR This disease is beyond my practice; yet I have known those which have walked in their sleep, who have died holily in their beds.

LADY MACBETH Wash your hands, put on your night-gown, look not so pale: I tell you yet again Banquo's buried; he cannot come out on's grave.   60

DOCTOR Even so?

LADY MACBETH To bed, to bed—there's knocking at the gate—come, come, come, come, give me your hand— what's done, cannot be undone. To bed, to bed, to bed.   *Exit*

DOCTOR Will she go now to bed?

GENTLEWOMAN Directly.

65 *Exit*] F (*Exit Lady.*)

40–1 **The Thane … now** This sounds like a tag from an old song, though presumably none existed.

53 **for … body** The comment is clear overall, but the exact sense is hard to catch: 'dignity' meant 'worth' as well as up-rightness, and the form of the phrase suggests 'for the sake of', with 'dignity' making a stronger impact than 'sake'.

56 **practice** professional skill, art

61 **on's** of his

65 **what's done, cannot be undone** Proverbial, Dent T200.

DOCTOR

Foul whisp'rings are abroad: unnatural deeds
Do breed unnatural troubles; infected minds   70
To their deaf pillows will discharge their secrets:
More needs she the divine than the physician—
God, God forgive us all. Look after her,
Remove from her the means of all annoyance,
And still keep eyes upon her. So good night,
My mind she has mated, and amazed my sight.
I think, but dare not speak.

GENTLEWOMAN   Good night, good doctor.

Exeunt

5.2   *Drum and colours.*
       Enter Menteith, Caithness, Angus, Lennox, Soldiers

MENTEITH

The English power is near, led on by Malcolm,
His uncle Seyward, and the good Macduff.
Revenges burn in them: for their dear causes
Would to the bleeding and the grim alarm
Excite the mortified man.

ANGUS   Near Birnam Wood

---

74 **annoyance** injury (from annoy, *v.* 3;
    OED does not give this sense for the noun)
76 **mated** baffled
    amazed bewildered, astounded
77 Proverb 'one may think that dare not
    speak', Dent T220.
5.2.0.2 **Menteith, Caithness, Angus** Shakes-
    peare seems to have drawn these names
    from Holinshed's list of those who were
    made earls by Malcolm (see 5.7.93–4);
    Holinshed also mentions Atholl and
    Murrey, so presumably the three names
    used give an index of the number of
    actors expected to be available for this
    scene; Angus speaks in Act 1, but then
    fades out, and none of the three is ever
    identified in dialogue.
2 **uncle** Holinshed states that Seyward's
    daughter was Duncan's wife, which
    would make Seyward Malcolm's grand-
    father; Muir points out that as
    Shakespeare transformed Holinshed's
    Duncan, a relatively young and feeble
    king, into the revered 'good old man', so

he apparently adjusted the relationship
to fit. See Introduction, pp. 68–9.
3–5 The multiple suggestions cannot be
    fully disentangled since several senses of
    each word interact with several of the
    others. The burning fever of revenge can
    only be cured by letting blood; a dead
    man's wounds are supposed to bleed in
    the presence of his murderer; the most
    insensitive man would rally to these
    causes; even the dead would quicken in
    response to such an alarm; etc.
3 **dear** 1. close to their hearts; 2. honour-
    able; 3. dire (*OED*, *a.*² 2)
    causes 1. grounds for revenge; 2. stir-
    rings of others to action; 3. diseases
    (revenge is a burning fever needing cure;
    see l. 15)
4 **bleeding** 1. (adjective) bloody (war); 2.
    (noun) medical blood-letting (to cure
    fever); 3. bleeding wounds
    alarm call to war
5 **mortified** 1. dead; 2. dulled, insensible (as
    in death)

Shall we well meet them, that way are they coming.
CAITHNESS
Who knows if Donalbain be with his brother?
LENNOX
For certain sir, he is not: I have a file
Of all the gentry; there is Seyward's son,
And many unrough youths that even now   10
Protest their first of manhood.
MENTEITH   What does the tyrant?
CAITHNESS
Great Dunsinan he strongly fortifies;
Some say he's mad; others, that lesser hate him,
Do call it valiant fury, but for certain
He cannot buckle his distempered cause
Within the belt of rule.
ANGUS   Now does he feel
His secret murders sticking on his hands,
Now minutely revolts upbraid his faith-breach:
Those he commands move only in command,
Nothing in love; now does he feel his title   20
Hang loose about him, like a giant's robe
Upon a dwarfish thief.
MENTEITH   Who then shall blame
His pestered senses to recoil, and start,
When all that is within him does condemn
Itself for being there?
CAITHNESS   Well, march we on,
To give obedience where 'tis truly owed;
Meet we the med'cine of the sickly weal
And with him pour we in our country's purge,
Each drop of us.
LENNOX   Or so much as it needs,

5.2.10 unrough] F (vnruffe); unruff'd POPE

8 **file** list
10 **unrough** beardless
11 **Protest** declare, show forth
15 **distempered** 1. disordered; 2. diseased
   cause See l. 3.
17 **sticking** as with blood
18 **minutely** i.e. every minute (*OED, adv.*²)

23 **pestered** plagued
   to recoil for recoiling
26 **owed** i.e. to Malcolm
28 **purge** medicine, to purge the system
29 **Each drop of us** 1. each one of us as a drop
   of medicine; 2. each death of us; 3. each
   drop of our blood

To dew the sovereign flower and drown the weeds: 30
Make we our march towards Birnam.

>   *Exeunt marching*

5.3   *Enter Macbeth, Doctor, and Attendants*

MACBETH

Bring me no more reports, let them fly all:
Till Birnam Wood remove to Dunsinan
I cannot taint with fear. What's the boy Malcolm?
Was he not born of woman? The spirits that know
All mortal consequences have pronounced me thus:
'Fear not, Macbeth, no man that's born of woman
Shall e'er have power upon thee'. Then fly false
        thanes,
And mingle with the English epicures;
The mind I sway by, and the heart I bear,
Shall never sag with doubt, nor shake with fear.   10

>   *Enter Servant*

The Devil damn thee black, thou cream-faced loon:
Where got'st thou that goose-look?

SERVANT

There is ten thousand—

MACBETH   Geese, villain?

SERVANT   Soldiers, sir.

5.3.5 consequences] F; consequence SINGER

30 **sovereign** 1. excellent; 2. medically effi-
cacious; 3. royal (i.e. Malcolm)
5.3.0 At some point during this scene Macbeth's armour must be brought on stage:
it could be here, by the attendants; or it
could be set behind the curtain of the
discovery space, which Seyton can draw
back; or Seyton may leave the stage to
fetch it after l. 33 (or 35). Evidently
Seyton (obeying l. 57) carries it off at the
end of the scene.
3 **taint** lose courage, become weak
(*OED* C.3)
4 **spirits** the 'masters' of the Weïrd Sisters
5 **mortal consequences** 1. outcomes for all
mortals; 2. death (i.e. 'when we will die'),
because death is the consequence of
mortality. The line is a hexameter, rhythmically damaged by emendation.
8 **English epicures** The English are fond of
pronouncing southern Europeans epi-

cures; apparently the Scots said the same
of their southern neighbours: G. K.
Hunter says the belief is traditional;
Holinshed says that the 'Scottish people
before had no knowledge nor understanding of fine fare or riotous surfeit ...
those superfluities which came ... with
the *Englishmen*' (see Introduction, p. 71).
9 **sway** 1. govern (in man the microcosm,
mind should rule as king rules realm);
2. move back and forth (as his mind is
about to)
11 **loon** Originally rogue or idler, it was also
used for anyone of low condition (the
opposite of kings) in social rank, or in
intelligence, or in age: hence, fool or
clown, and boy or lad. All senses are
relevant here.
12 **goose** white; simpleton; terrified (as in
goose-flesh)

MACBETH

Go prick thy face, and over-red thy fear,
Thou lily-livered boy. What soldiers, patch?
Death of thy soul, those linen cheeks of thine
Are counsellors to fear. What soldiers, whey-face?

SERVANT

The English force, so please you.

MACBETH

Take thy face hence.—   *Exit Servant*
              Seyton—I am sick at heart
When I behold—Seyton, I say—this push   20
Will cheer me ever, or dis-seat me now.
I have lived long enough: my way of life
Is fall'n into the sere, the yellow leaf,
And that which should accompany old age,
As honour, love, obedience, troops of friends,
I must not look to have—but in their stead
Curses, not loud but deep, mouth-honour, breath
Which the poor heart would fain deny, and dare not.—
Seyton?

              Enter Seyton

19 *Exit Servant*] *not in* F   21 cheer] F (cheere); chair SINGER 1856 (*conj.* Percy)   dis-seat]
STEEVENS; dis-eate F1; disease F2   22 way] F; May STEEVENS (*conj.* Johnson)

14 **prick … fear** 1. literally, use your blood
to disguise your fear; 2. make your face
tingle and so blush; 3. 'prick' in the sense
of spur, i.e. pull yourself together and
recover your normal complexion

15 **lily-livered boy** See proverbs 'as white as
a lily', 'a white-livered fellow', Dent
L296, F180.
patch domestic fool, clown, dolt (*OED sb.*
2, from Italian *pazzo*, a fool, possibly
conflated with the patched costume of
the fool)

16 **linen cheeks** Proverb 'as white as linen',
Dent L306.1.

20 **behold** Macbeth seems to break off, leaving several implications: 1. see frightened
faces; 2. see that (this push etc.); 3.
seeing, contemplating at all, terrifies me
now.
push thrust, attack (a strong sense, close
to 'offensive' in recent military jargon)

21 **cheer** 1. encourage; 2. enthrone (punning with 'chair')
dis-seat F's 'dis-eate' is probably a misprint since its only possible sense, 'vomit

my life out', is too strained; F2's 'disease'
is a plausible guess and may even be a
pun, but Steevens's 'dis-seat', meaning
unseat from throne and life, is much
more probable in the context.

22 **way of life** The phrase has, oddly, been
argued about and emended; life's path
(way) is from birth to death via youth and
age, as the seasons go from spring to
winter. The modern sense of 'way of life',
'manner of living', may also be involved,
hence 'fallen' (from high to low).

23 **sere** dry, desiccated, sapless
yellow leaf i.e. autumn

25 **As** such as

27 **mouth-honour** i.e. lip-service
breath whisper, utterance, speech
(*OED* 9)

29 **Seyton** G. R. French, *Shakespeareana
Genealogica* (1869), claimed that the
Setons had been and still were hereditary
armour-bearers to the kings of Scotland.
Muir scorns a suggestion that there
might be a pun on 'Satan', but it is
very possible: Marlowe has Lightborne

SEYTON   What's your gracious pleasure?

MACBETH   What news more?

SEYTON

All is confirmed, my lord, which was reported.   30

MACBETH

I'll fight, till from my bones my flesh be hacked.

Give me my armour.

SEYTON   'Tis not needed yet.

MACBETH I'll put it on.

Send out more horses, skirr the country round,

Hang those that talk of fear. Give me mine armour—

How does your patient, doctor?

DOCTOR   Not so sick, my lord,

As she is troubled with thick-coming fancies

That keep her from her rest.

MACBETH   Cure her of that:

Canst thou not minister to a mind diseased,

Pluck from the memory a rooted sorrow,   40

Raze out the written troubles of the brain,

And with some sweet oblivious antidote

Cleanse the stuffed bosom of that perilous stuff

Which weighs upon the heart?

DOCTOR   Therein the patient

Must minister to himself.

MACBETH

Throw physic to the dogs, I'll none of it.

Come, put mine armour on; give me my staff—

Seyton, send out.—Doctor, the thanes fly from me.—

34 more] F (moe)   38 Cure her] F2; Cure F1   43 stuffed ... stuff] F (stufft ... stuffe);
stuffed ... grief COLLIER MS; charged ... stuff WILSON; fraught ... stuff OXFORD

(Lucifer) in Act 5 of *Edward II*, and
Shakespeare has Mercadé (Macabré, sup-
posed author of the 'Dance of Death') in
Act 5 of *LLL*.

34 **more** F's 'moe' was a common form of
'more'
   skirr run rapidly, here to scour it
   thoroughly
38 **Cure her** F2's insertion is obviously
needed; the omission in F1 was doubtless
a slip.
41 **written** inscribed, carved (as Mary I said
Calais was on her heart)
42 **oblivious** causing oblivion

43 **stuffed ... stuff** Editors have worried
about the repetition, but 'stuff' is used in
two distinct senses: 1. as a verb, 'clogged'
(*OED* 12, specifically applied to bodily
organs); 2. as a noun, 'matter', whether
solid or liquid, here evidently a poison
which requires an antidote. All proposed
emendations seem to me inferior: the
repetition of 'stuff' insists on the dreadful
anonymity of morbid oppression.
47 **staff** either a weapon (stick, lance, etc.),
or a staff of office (e.g. marshal's baton,
first referred to in 1590)
48 **send out** See l. 34.

Come sir, dispatch.—If thou couldst, doctor, cast
The water of my land, find her disease, 50
And purge it to a sound and pristine health,
I would applaud thee to the very echo
That should applaud again.—Pull't off I say.—
What rhubarb, senna, or what purgative drug
Would scour these English hence? Hear'st thou of
     them?

DOCTOR
Ay my good lord: your royal preparation
Makes us hear something.

MACBETH (*to Seyton*)   Bring it after me—
I will not be afraid of death and bane
Till Birnam forest come to Dunsinan.
                    ⌈*Exeunt Macbeth and Seyton*⌉

DOCTOR
Were I from Dunsinan away and clear, 60
Profit again should hardly draw me here.   *Exit*

5.4   *Drum and colours.*
       Enter Malcolm, Seyward, Macduff, Seyward's son,
       Menteith, Caithness, Angus, and Soldiers, marching

MALCOLM
Cousins, I hope the days are near at hand

51 pristine] F (pristiue)   54 senna] F4; Cyme F1; Coeny F2, F3   59.1 *Exeunt ... Seyton*]
not in F   61 *Exit*] F (*Exeunt*)
5.4.0.1–2] F; ... *and Lennox, Ross* MALONE

49 **dispatch** make haste, presumably to put
the armour on; its other sense connects it
with 'send out' in l. 48
49–50 **cast | The water** 'cast' in the sense
of 'reckon, calculate' (*OED* VI), i.e. ana-
lyse the urine, a common process of
medical diagnosis
53 **Pull't off** i.e. the armour: he has changed
his mind (see 'Bring it after me' in l. 57)
54 **rhubarb** the bitter root of Chinese rhubarb, used medicinally as a purge
   senna 'cyme', F1, is a technical botanical
term without medical significance (it was
also sometimes used for the tender buds
of any plant); it is generally assumed to
be here a misreading of 'cynne', a spelling
of 'cœny' or 'sene' which were common
names (usually disyllabic) for 'senna'.
Senna is both purgative and emetic.
Though hypermetrical, the disyllable

seems (to me) more satisfactory rhythmically.
58 **bane** 'murderer', 'murder', 'violent
death', 'ruin' are all possible senses here
5.4.0.2–3 It seems plausible to add, as most
editors do, Lennox and Ross here, though
they are not needed. The complex doubling required for a relatively small company to supply two armies probably
explains their absence: Lennox is essential to 5.2 (where he finally comes out
against Macbeth) but is not named after
that; Ross is not named in Act 5 until
5.7.64.4–5. There are five significant
new roles to be cast: Gentlewoman,
Doctor, Seyton, old and young Seyward;
the actors of Lennox and Ross may have
taken one or more each of these. See
Introduction, pp. 84–6.

That chambers will be safe.
MENTEITH   We doubt it nothing.
SEYWARD
What wood is this before us?
MENTEITH   The wood of Birnam.
MALCOLM
Let every soldier hew him down a bough
And bear't before him, thereby shall we shadow
The numbers of our host, and make discovery
Err in report of us.
SOLDIER   It shall be done.
SEYWARD
We learn no other but the confident tyrant
Keeps still in Dunsinan, and will endure
Our setting down before't.
MALCOLM   'Tis his main hope: 10
For where there is advantage to be given,
Both more and less have given him the revolt,
And none serve with him but constrainèd things,
Whose hearts are absent too.
MACDUFF   Let our just censures
Attend the true event, and put we on
Industrious soldiership.
SEYWARD   The time approaches
That will with due decision make us know
What we shall say we have, and what we owe;
Thoughts speculative their unsure hopes relate,
But certain issue, strokes must arbitrate—   20
Towards which, advance the war.   *Exeunt marching*

11 given] F; gone CAPELL

2 **chambers** domestic rooms, i.e. private
rooms, especially bed-chambers (where
Duncan died)
6 **discovery** reconnaissance
10 **Our setting down before't** i.e. encamping
for a siege
11–12 These lines have been much emended, but not much improved. I take
the sense to be 'Where they could have
given Macbeth the advantage (by staying
loyal) people of all social levels have
chosen to rebel against him'. Muir, like
many others, accepted Johnson's 'gone'
for 'given', but not his 'a vantage' for 'advantage'; he therefore glossed 'advantage'
as 'opportunity', for which there is no
warrant. Muir suggested that the compositor picked up 'given' from the line
below, but it seems to me that the repetition stresses the free choice of the people,
and also that 'give advantage' is a familiar
phrase where the alternatives are not.
12 **more and less** i.e. in social degree
14 **just censures** final judgement
18 **owe** i.e. to uncertain factors
19 i.e. speculation tells of uncertainty (Seyward, like Macduff, wants to put the
matter to the test of battle)

5.5   *Enter Macbeth, Seyton, and Soldiers, with drum and*
        colours

MACBETH

Hang out our banners on the outward walls,
The cry is still, 'They come': our castle's strength
Will laugh a siege to scorn; here let them lie
Till famine and the ague eat them up—
Were they not forced with those that should be ours,
We might have met them dareful, beard to beard,
And beat them backward home. What is that noise?

A cry within of women

SEYTON

It is the cry of women, my good lord.   ⌐*Exit*⌐

MACBETH

I have almost forgot the taste of fears:
The time has been, my senses would have cooled   10
To hear a night-shriek, and my fell of hair
Would at a dismal treatise rouse and stir
As life were in't. I have supped full with horrors,
Direness familiar to my slaughterous thoughts
Cannot once start me.

⌐*Enter Seyton*⌐

                    Wherefore was that cry?

SEYTON The Queen, my lord, is dead.

MACBETH She should have died hereafter;
There would have been a time for such a word—
Tomorrow, and tomorrow, and tomorrow,

5.5.2 'They come'] F (∧ they come ∧)   8 *Exit*] *not in* F   15 *Enter Seyton*] *not in* F

5.5.5 forced reinforced (with a possible implication that the rebels were conscripted
rather than volunteers as Malcolm had
claimed)
6 See proverb 'to meet in the beard' Dent
B143.1.
  dareful boldly
8 **cry of women** for the Queen. Muir
comments that 'Lady Macbeth did not die
a natural death', but Malcolm's insinua-
tion in 5.7.100–1 is not supported here.
The cry may be simply of dismay, but the
plural 'women' might imply the Irish
custom of keening, common to many
cultures though I do not know that it was
ever associated with the Scots. It seems
most likely that Seyton goes off to invest-

igate the cry but it is possible that someone enters to tell him, or even that in
l. 16 he simply realizes what has happened. 5.7.100–1 leave the manner of
her death to rumour only.
11 **fell** entire hair, usually with skin (but
used of the fleece of sheep)
12 **treatise** story (whether written or
spoken); the word was used of any
literary work, not specifically of a disser-
tation
17 The short line presumably follows a long
pause rather than precedes one.
She … hereafter 1. she should have lived
a full span; 2. she should have died at a
time when we had leisure to mourn
18 **word** i.e. death

Creeps in this petty pace from day to day,   20
To the last syllable of recorded time;
And all our yesterdays have lighted fools
The way to dusty death. Out, out, brief candle,
Life's but a walking shadow, a poor player
That struts and frets his hour upon the stage,
And then is heard no more. It is a tale
Told by an idiot, full of sound and fury
Signifying nothing.
     Enter a Messenger
Thou com'st to use thy tongue—thy story quickly.
MESSENGER Gracious my lord,   30
I should report that which I say I saw,
But know not how to do it.
MACBETH   Well, say, sir.
MESSENGER
As I did stand my watch upon the hill
I looked toward Birnam, and anon methought
The wood began to move.
MACBETH   Liar, and slave.
MESSENGER
Let me endure your wrath, if't be not so—
Within this three mile may you see it coming.
I say, a moving grove.
MACBETH   If thou speak'st false,
Upon the next tree shall thou hang alive

39 shall] F1; shalt F2

20 **in this petty pace** 1. at this meaningless pace; 2. in this petty way; 3. in this narrow passage—'pace' in the early seventeenth century was used for a narrow passage, pass, or strait, which would make sense here and explain the use of 'in' rather than 'at'

21 A line too complex in its resonances for commentary to be anything but reductive: 'syllable' had the sense of the least trace of *anything*, as well as the least part of a word; 'recorded time' must surely allude to the recording angel who reports all the doings of souls to the Last Judgement when time has stopped and all is changed eternally.

22–4 **fools ... shadow ... player** All words for actors, though used here in general senses as well; see *Dream* 5.1.210, 'The best in this kind [actors] are but shadows'.

23–4 **dusty death ... walking shadow** There are several biblical contexts for 'dust' and 'shadow', many of them used in the Prayer Book Order for the Burial of the Dead, e.g. 'Man that is born of a woman hath but a short time to live, and is full of misery ... he fleeth as it were a shadow', 'earth to earth, ashes to ashes, dust to dust'. See proverb 'life is a shadow', Dent L249.1.

24 **poor** 1. unfortunate; 2. feeble

24–5 **player ... stage** Proverb 'this world's a stage and every man plays his part' Dent W882.

26–7 **a tale | Told** Psalm 90, prescribed for the burial service, has 'we bring our years to an end, as it were a tale that is told'.

Till famine cling thee; if thy speech be sooth, 40
I care not if thou dost for me as much.
I pull in resolution, and begin
To doubt th'equivocation of the fiend
That lies like truth. 'Fear not, till Birnam Wood
Do come to Dunsinan', and now a wood
Comes toward Dunsinan.—Arm, arm, and out—
If this which he avouches does appear,
There is nor flying hence, nor tarrying here.
I 'gin to be aweary of the sun
And wish th'estate o'th' world were now undone. 50
Ring the alarum bell, blow wind, come wrack,
At least we'll die with harness on our back. *Exeunt*

5.6 *Drum and colours.*
Enter Malcolm, Seyward, Macduff, and their army,
with boughs

MALCOLM
Now near enough: your leafy screens throw down,
And show like those you are.—You, worthy uncle,
Shall with my cousin your right noble son
Lead our first battle. Worthy Macduff and we
Shall take upon's what else remains to do,
According to our order.

SEYWARD   Fare you well—
Do we but find the tyrant's power tonight,
Let us be beaten if we cannot fight.

MACDUFF
Make all our trumpets speak, give them all breath,
Those clamorous harbingers of blood, and death. 10
                                        Exeunt

   Alarums continued

42 pull] F; pall HUDSON (*conj.* Johnson)   44–5 'Fear not ... Dunsinan'] F (∧Feare not ...
Dunsinane∧)

40 **cling** shrink, wither (*OED* 2)
   sooth truth
42 **pull in resolution** 'pull in' is used for
   reining in or checking a horse; 'resolution' for conviction, firmness of purpose;
   the text gives a better sense than the
   common emendation to 'pall', though
   that may still be a punning suggestion.
   OED, 'pull', *v.* 25d; this line is the sole
   recorded use in this sense before 1780.

50 **estate** in several senses, especially 'con-
   stitution', 'state'
51 **wrack** disaster, ruin, vengeance, etc.
52 **harness** Used of military equipment and
   of armour, as well as of horse gear.
5.6.4 **battle** one of the main divisions of an
   army in battle array
6 **order** disposition, arrangement
7 **power** army

5.7   *Enter Macbeth*

MACBETH

They have tied me to a stake, I cannot fly,
But bear-like I must fight the course. What's he
That was not born of woman? Such a one
Am I to fear, or none.

   Enter Young Seyward

YOUNG SEYWARD

What is thy name?

MACBETH   Thou'lt be afraid to hear it.

YOUNG SEYWARD

No: though thou call'st thyself a hotter name
Than any is in Hell.

MACBETH   My name's Macbeth.

YOUNG SEYWARD

The Devil himself could not pronounce a title
More hateful to mine ear.

MACBETH   No, nor more fearful.

YOUNG SEYWARD

Thou liest abhorrèd tyrant, with my sword   10
I'll prove the lie thou speak'st.

   Fight, and Young Seyward slain

MACBETH Thou wast born of woman;
But swords I smile at, weapons laugh to scorn,
Brandished by man that's of a woman born.   *Exit*

   Alarums.

   Enter Macduff

5.7 F gives this as all one scene whereas most editors have introduced two further divisions, at ll. 30 and 64: the first because Malcolm is invited to 'enter the castle', the second because (according to Muir) there's no reason why he should leave it again. But in the Globe Theatre there was no scenery, and if for a moment the tiring-house represents the castle, Malcolm must leave it to re-enter the stage. The fact is that Scenes 2–6 represent the preparations for battle when the two armies must be understood to be in separate locations (and Malcolm is marching), Scene 7 is the battle itself which is continuous, where a number of incidents happen before our eyes and the rest is off-stage. The Folio arrangement is therefore entirely rational, and any other

forgets the reality of the theatre for an impossible series of mini-scenes designated 'Another part of the field'.

1 Proverb 'to be bound to a stake', Dent S813.1.

2 **bear-like ... course** An allusion to bear-baiting where the bear was tied to a stake in the middle of the pit and attacked by several small dogs at once. 'course' was used of encounters in battle as well as of sporting events.

11.1 Young Seyward's body must presumably be removed from the stage before l. 24 when his father enters. It is frequently a problem in Jacobean plays to get rid of the bodies, and nothing seems to be known about how this was done at the time. (Oxford directs Macbeth to remove it after l. 14.)

MACDUFF

That way the noise is: tyrant show thy face;
If thou be'st slain, and with no stroke of mine,
My wife and children's ghosts will haunt me still—
I cannot strike at wretched kerns, whose arms
Are hired to bear their staves; either thou Macbeth,
Or else my sword with an unbattered edge   20
I sheathe again undeeded. There thou shouldst be,
By this great clatter one of greatest note
Seems bruited. Let me find him, fortune,
And more I beg not.   *Exit*

    Alarums.
    Enter Malcolm and Seyward

SEYWARD

This way my lord, the castle's gently rendered:
The tyrant's people on both sides do fight,
The noble thanes do bravely in the war,
The day almost itself professes yours,
And little is to do.

MALCOLM   We have met with foes
That strike beside us.

SEYWARD   Enter, sir, the castle.   *Exeunt*   30

    Alarum.
    Enter Macbeth

MACBETH

Why should I play the Roman fool, and die
On mine own sword? Whiles I see lives, the gashes
Do better upon them.

    Enter Macduff

---

18 **kerns** rustic foot-soldiers (see 1.2.13), here evidently very reluctantly conscripted

18–19 **arms … staves** 'arms' is used both literally, and to stand for the men, hired solely as arms to bear 'staves', mere poles, the crudest form of weapon (also used of the shaft of a spear)

21 **undeeded** with no deeds committed (apparently a coinage here)

22 **note** 1. notability; 2. musical note (ironically with 'clatter'—no trumpet is called for)

23 **bruited** 1. announced; 2. sounded (loudly: the French 'bruire' meant to roar)

25 **rendered** handed over, surrendered

30 **strike beside us** 1. deliberately hit to miss (see *Richard Duke of York* (3 *Henry VI*) 2.1.130–2, 'Our soldiers … | Fell gently down, as if they struck their friends'); 2. turn and fight beside us

30.2 *Enter Macbeth* See note to beginning of scene.

31 **the Roman fool** Referring to the dignity in Roman culture of ending a life defeated by suicide, falling on one's own sword; Shakespeare dramatized two such, Brutus and Antony, both of whom failed to do it for themselves and ended in fumbled efforts to get their officers to do it for them.

MACDUFF   Turn hell-hound, turn.

MACBETH

Of all men else I have avoided thee;
But get thee back, my soul is too much charged
With blood of thine already.

MACDUFF   I have no words,
My voice is in my sword, thou bloodier villain
Than terms can give thee out.

        Fight. Alarum

MACBETH   Thou losest labour—

As easy mayst thou the intrenchant air
With thy keen sword impress, as make me bleed: 40
Let fall thy blade on vulnerable crests,
I bear a charmèd life, which must not yield
To one of woman born.

MACDUFF   Despair thy charm,
And let the angel whom thou still hast served
Tell thee, Macduff was from his mother's womb
Untimely ripped.

MACBETH

Accursèd be that tongue that tells me so,
For it hath cowed my better part of man;
And be these juggling fiends no more believed
That palter with us in a double sense,   50
That keep the word of promise to our ear
And break it to our hope. I'll not fight with thee.

MACDUFF Then yield thee, coward,
And live to be the show and gaze o'th' time.
We'll have thee, as our rarer monsters are,

37–8 **bloodier ... | Than** i.e. villain who is bloodier than

39 **intrenchant** uncuttable; literally 'un-cutting', used in a passive sense of a verb that is usually active, so meaning 'not being cut'—another negative/passive coinage (see l. 21)—because the air closes back round the sword which leaves no mark.

40 **impress** imprint, mark by pressure (the word was also used for forcing military service and ll. 18–19 may influence its use here)

41 **crests** plumes on a helmet, hence helmet or head

44 **angel** i.e. fallen angel, the Devil; or, Macbeth's 'bad angel'

46 **Untimely ripped** Prematurely, by sur-gery—presumably because of the sick-ness or death of his mother; the short line suggests a pause before Macbeth re-sponds.

48 **better part of man** The phrase was used of the spirit or soul (see Sonnet 74); here it evidently includes the courageous spirit of manliness, Macbeth's recurrent con-cern.

50 **palter** shuffle, equivocate

55–6 Monsters were exhibited, as they still were until recently in fairground freak

Painted upon a pole, and underwrit
'Here may you see the tyrant.'
MACBETH   I will not yield
To kiss the ground before young Malcolm's feet,
And to be baited with the rabble's curse.
Though Birnam Wood be come to Dunsinan,   60
And thou opposed, being of no woman born,
Yet I will try the last. Before my body
I throw my warlike shield: lay on Macduff,
And damned be him that first cries 'Hold, enough!'
                         Exeunt fighting

    Alarums.
    Re-enter fighting, and Macbeth slain
           ⌜*Exit Macduff with Macbeth's body*⌝
    Retreat, and flourish.
    Enter with drum and colours Malcolm, Seyward,
    Ross, Thanes, and Soldiers

MALCOLM
I would the friends we miss were safe arrived.

SEYWARD
Some must go off; and yet by these I see
So great a day as this is cheaply bought.

MALCOLM
Macduff is missing, and your noble son.

ROSS
Your son, my lord, has paid a soldier's debt;

5.7.57 'Here ... tyrant.'] F (ˌHeere ... Tyrant.ˌ)   64.3 *Re-enter ... slain*] F (*Enter*), WILSON ; *not in* POPE   64.4 *Exit Macduff ... body*] OXFORD ; *not in* F

shows, inside a tent or booth with a painted sign outside to attract customers; see *Tempest* 2.2.27–30, 'Were I in England now ... and had but this fish [Caliban] painted, not a holiday-fool there but would give a piece of silver'.

58 **kiss the ground** Proverbial, Dent D651 (cf. 'lick the dust').

61 **opposed** set against (for fighting), as in *Hamlet* 3.1.61–2: 'Or to take arms against a sea of troubles, | And, by opposing, end them'.

64.1–3 F's direction suggests that the stage is left bare for a short while, an unusual effect used a few years later by Middleton in *The Changeling* (1622), 3.1.10, where

De Flores is leading Alonso to his death, with the direction '*Exeunt at one door and enter at the other*'. F has no direction for Macduff's exit with the body; this may be simple omission, or perhaps Macbeth was slain in the discovery space and curtains drawn before Malcolm's entry; otherwise Macduff must kill on the main stage and lug the body through one of the doors, Malcolm entering by the other. (See also note at beginning of the scene.)

66 **go off** i.e. die; Muir suggests plausibly a theatrical metaphor 'exit from life's stage'.

69 **paid a soldier's debt** See proverbs 'Death pays all debts', 'to pay one's debt to nature', Dent D148, D168.

He only lived but till he was a man,   70
The which no sooner had his prowess confirmed
In the unshrinking station where he fought,
But like a man he died.
SEYWARD   Then he is dead?
ROSS
Ay, and brought off the field; your cause of sorrow
Must not be measured by his worth, for then
It hath no end.
SEYWARD   Had he his hurts before?
ROSS
Ay, on the front.
SEYWARD   Why then, God's soldier be he:
Had I as many sons as I have hairs,
I would not wish them to a fairer death—
And so his knell is knolled.
MALCOLM   He's worth more sorrow,   80
And that I'll spend for him.
SEYWARD   He's worth no more,
They say he parted well, and paid his score,
And so God be with him. Here comes newer comfort.
          Enter Macduff, with Macbeth's head
MACDUFF (*to Malcolm*)
Hail King, for so thou art. Behold where stands
Th'usurper's cursèd head: the time is free;
I see thee compassed with thy kingdom's pearl,
That speak my salutation in their minds,
Whose voices I desire aloud with mine:
Hail, King of Scotland.

---

72 **station** stand; *OED* I.3 gives this sense,
which here doubles with the more usual
'standing-place' in which Seyward made
his unshrinking stand.
77 **front** forehead, sometimes generalized to
the whole face
78 Referring to proverb 'as many as there
are hairs on the head' (Dent H30), but
punning with 'heirs'.
83.1 Presumably the head is on the end of
his lance (see l. 84 'where stands ...'):
Holinshed says 'he set it upon a pole', as a
gruesome version of the painted image in

ll. 55–6. I take it that a life-mask of
Burbage would supply the necessary life-
likeness, painted and dripping blood;
compare the interest in totally life-like art
attributed to Giulio Romano's statue of
Hermione at the end of *Winter's Tale.*
85 **the time is free** i.e. our country is free in
our time (see 1.5.62)
86 **compassed ... pearl** 1. surrounded by the
thanes, the aristocratic jewels of the
country; 2. compassed with the royal
crown of Scotland

ALL   Hail, King of Scotland.

  Flourish

MALCOLM

We shall not spend a large expense of time   90
Before we reckon with your several loves,
And make us even with you. My thanes and kinsmen,
Henceforth be earls, the first that ever Scotland
In such an honour named. What's more to do
Which would be planted newly with the time,
As calling home our exiled friends abroad
That fled the snares of watchful tyranny,
Producing forth the cruel ministers
Of this dead butcher, and his fiend-like Queen
Who, as 'tis thought, by self and violent hands 100
Took off her life—this, and what needful else
That calls upon us, by the grace of Grace
We will perform in measure, time, and place:
So thanks to all at once, and to each one,
Whom we invite to see us crowned at Scone.

  Flourish.   *Exeunt omnes*

90–105 Malcolm's acceptance speech is solely concerned with securing his own power and makes no reference to the suffering countrymen that Macduff spoke of in 4.3.

91 **we** Malcolm immediately adopts the royal form.
 reckon with settle accounts with

93–4 **earls ... named** Holinshed recounts this. James I was notorious for ennobling so many of his Scottish followers on his first ascending the English throne; the actors and several dramatists had been both censored and imprisoned for satiric references to new Scottish peers about 1605, notably Chapman, Marston, and Jonson.

100–1 **self ... life** Lady Macbeth's suicide is not suggested in Holinshed; it is given here only as rumour and was not mentioned in 5.5, where it would be more natural to assume that despair led to loss of the will to live. The ambiguity is apt: suicide was a crime as well as a sin, and Malcolm would naturally attribute both to her.

102 **grace of Grace** the grace of God, apostrophized as the essence of graciousness

LINEATION

THE following aims at complete collation of all lines in the Folio text of *Macbeth* where doubt has arisen, or could reasonably arise, as to their proper arrangement, whether as verse or prose. The distinction is not easily drawn between merely different typographical conventions, and possible sources of confusion: both Folio compositors take runover words (where a line was too long for their column) to the beginning of a new line and give it an initial capital; both also, where a line is divided between speakers, give each speaker's contribution a fresh line, where modern texts conventionally space the words along the line to make its verse structure apparent. The normal occurrence of both these phenomena is so frequent that collation is impossible, and would in any case obscure the serious problems that the text presents. Other habits which do not normally cause any problem to an editor, but which might occasionally be held to have had some other origin than typographical convention, have been collated. The commonest are: 1. where two lines are divided between two speakers in four half-lines, such that the second speaker has two halves, that is often set out as if it were a single full line (thus saving space; manuscripts show the same habit)—e.g. 4.3.173–4, which appears in F as:

> Dying, or ere they sicken.
> *Macd.* Oh relation; too nice, and yet too true.
> *Malc.* What's the newest griefe?

(the same principle can be seen at 3.4.12–16 in a more confusing form). 2. Where the first line for a new speaker was too long for the column (as it often was, because prefixes were indented), instead of simply running over the extra words the line is often printed as two half-lines, e.g. at 3.4.123, Macbeth's words appear in F as:

> *Macb.* It will have blood they say:
> Blood will have blood;

It looks as though the decisive syntactical caesura has influenced the division of the line (which could not have been printed as one); and the compositor has emphasized this with heavy punctuation. 3. Shakespeare frequently ended a speech with a half-line, much more rarely began with one (e.g. 2.3.71–2, 2.4.32–3);[1] this occasionally leaves a gap, but

---

[1] See Fredson Bowers, 'Establishing Shakespeare's Text: Notes on Short Lines and the Problem of Verse Division', *Studies in Bibliography*, 33 (1980), 74–130 (especially 79 and 86).

usually the single half-line functions rhythmically both as end of one speech and beginning of the next, and no break is felt.

These examples are mostly from the work of Compositor B, but they occur with roughly equal frequency in the work of both compositors. In other respects their behaviour was very different: B never allows these conventions to distort the general sequence of the verse, nor does he ever prefer syntax to verse in setting out his lines. A did both in generous measure; he seems to have had no respect at all for the rhythm of verse as such (his experience had probably been largely with setting prose texts: I have encountered the same unawareness in the derangement of lengthy verse quotations by modern compositors), with the result that he often fills out one line with words from the next, and so continues for several at a stretch that *look* like verse, but do not *sound* like it at all; his respect for syntax usually gets him straight after three or four lines, but occasionally the fact that Shakespeare's verse is so consistently run-on extends the confusion to as much as eight or ten (e.g. 2.2.2–7).

Compositor A has been convicted of similar villainy elsewhere;[1] it follows inescapably that, wherever possible, eccentricities in his setting out of lines should be corrected. This process began with Rowe in 1709 (but indirectly it had already been done in late seventeenth-century adaptations, notably by Davenant), and was far more thoroughly achieved by Pope in 1728. Pope worked with characteristic sensitivity and intelligence, and respected some forms of apparent irregularity which he had defended in the 'Essay on Criticism'; he knew that 'Late, very late, correctness grew our care', and that Shakespeare cared less for it than he did. None the less his judgement was inevitably affected by his own (and his contemporaries') taste: he preferred regularity if he could possibly arrange it; he preferred long lines to short ones (as in his own practice, derived from Dryden); and he preferred verse to prose. These prejudices have affected editorial practice ever since, and the process was elaborated by Pope's successors in the eighteenth and nineteenth centuries. In the twentieth century sympathy has moved back towards the rhetorical flexibility of Jacobean verse, and especially of dramatic verse, composed by ear for speech, not by counting syllables, and the editions of Dover Wilson (1947) and Kenneth Muir (1950) accepted a number of Folio readings which had been traditionally altered. But the effect of an editorial tradition which has its own history is still visible, and current thinking inclines to leave it in place unless a very strong case can be made for reverting to the Folio.

With Compositor A, as I have said, emendation is clearly right: in the

---

[1] I have been helped by two unpublished essays: G. Taylor, 'The Sins of Compositor A', and P. Werstine, 'Line Division in Shakespeare's Dramatic Verse: An Editorial Problem.'

first half of the play (up to the end of 3.3) which he set nearly all of, I have altered the lineation of approximately 40 out of the 50 passages where error has been suspected. For most of these (30 passages) a convincing reorganisation has been found in the past and accepted by the vast majority of editors; for the other ten I have preferred an arrangement of my own devising, either because there has been no consensus, or because I find accepted emendations more characteristic of their eighteenth-century originators than of early seventeenth-century dramaturgy.

There remain another ten passages where no solution has found general acceptance, or, if one has, it seems to me no more satisfactory than the Folio reading; usually these occur when there is no obvious reason of the kinds I have described why Compositor A should have perverted his copy, so that reconstruction of a superior version—if there ever was one—is impossible; this may be either because of the compositor's errors, or of a copyist's, or because Shakespeare was careless about tidying up his slips or alterations; or because of cuts in the text.

The fact is that Jacobean dramatic verse in general, and Shakespeare's in particular, was not consistently regular. Lines that seem superficially defective do not always seem so when spoken, either because the stresses are clear enough to carry a variable number of unstressed syllables, or because a strong stop demands a syllabic pause before the speaker can continue. One speaker may continue a speech without regard to another's exclamation, or the completion of a line may patently depend on a dramatic pause, or even an action; but it must be clearly realized that there is no consistency about this: many pauses, moves, etc., are certainly required which have, as it were, to be extra-metrical—the verse continues without regard to them. Verse rhythm is rarely precisely the same as dramatic rhythm; but sometimes it is, and Shakespeare's awareness of his language in action can never be overestimated.

Richard Flatter, in *Shakespeare's Producing Hand* (1948), argued whole-heartedly for this idea, with no understanding of the compositor's eccentricities. Most of his examples were drawn from *Macbeth*; many of them were immediately discredited, and subsequent attempts to extend his theories to other plays had little success, even in those quarto texts which were closest to Shakespeare's own manuscripts. The one exception is the Folio text of *Coriolanus* where again Compositor A was at work, and probably responsible. For that reason any such argument has to be treated with great caution. The best-known instance is Flatter's defence of the Captain's speech at 1.2.7–42 in Compositor A's arrangement, which he thought represented the broken utterance of a seriously wounded man. In fact, simple rearrangement shows it to be regular and rhetorically impressive in the manner of a Senecan messenger, and only the half-line 20 could possibly sustain Flatter's thesis until the last two lines (41–2) where indeed it seems likely that a pause in the middle of l. 41 represents the exhaustion which the speaker is expressing.

In short, Shakespeare's use of metrical gaps to signify pauses seems to have been, at most, very rare, and cannot be relied on to justify the bad work of Compositor A. But there are a very few contexts where it may be a valid explanation, as in l. 41. I have tentatively noted the possibility of others in the annotations below, but accepted very few indeed into the text (a similar sensitivity can be seen in 2.2.17, where a very rapid exchange of speakers is stressed by containing five units in a single verse line).

Compositor B's characteristics were quite different: he has been shown to have freely substituted his own words for his copy's in many contexts, but he does have a correct ear for verse, so that where (as here) we have no earlier printed version to check his performance, his verbal substitutions are virtually undetectable. Similarly, if he extended this to adjusting the lineation of his copy, we cannot know where. But he does have one habit of perversion, that of printing prose as if it were verse; and this, I think, explains the extraordinary muddle he made of 4.2.31–53 (most editors take at least part of this for prose), and, though this is not the usual practice, I take 3.4.1–8 as also prose. In both passages it seems likely that he did little more than attach initial capitals to the lines of his copy, for the rapid dialogue keeps largely to short units. Dramatic manuscripts, including the three pages of *Sir Thomas More* generally accepted as Shakespeare's, are very sparing with initial capitals for verse lines, so prose and verse can easily be confused. In addition, it was a common practice with Jacobean dramatists to make a gradual transition from prose to verse, via fragments of verse rhythm into a more regular flat verse which can then progressively develop a stronger rhythm (the first 39 lines of *Hamlet* are a striking example). A similar problem arises occasionally in Compositor A's stint, notably at 3.1.75–91 which I print as prose.

It follows that my practice in preparing the text has been, in Compositor A's half of the play, to realign doubtful passages wherever possible and to seek other explanations only where no straightforward solution seems to be possible; in Compositor B's stint, to accept good prose where it seems preferable to bad verse, but otherwise only to tidy up what are, in fact, merely typographical differences. Explanations for errors are tentatively offered in the notes; where none is given it is usually because none seems to be needed; where confusion has arisen simply because of the narrow column in the Folio, this is indicated by the single word 'space'. One practice which cannot be indicated typographically is where, in dialogue between two speakers, a single half-line serves both to complete the first speaker's incomplete line, and to be the first half of his next; in such cases it may be ambiguous whether the interjection is extra-metrical and the first speaker continues uninterrupted.

1.1–3.3 : Compositor A

1.1.9–11] SINGER 1856; *one line, prefix All.* F; *two lines, 2 Witch.* Padocke calls—anon | *All.* Fair … fair POPE (It's fairly obvious that this is at least two lines, crowded by the compositor, most probably by the copyist before him; since Graymalkin and Paddock are individual familiars they should be addressed by one witch each; giving 'Anon' to the third witch depends on the general rule that they speak in strict rotation.)

1.2.20] F (Various conjectures have been made to fill out this line, none has been accepted. It is either a pause for breath in a very elaborate sentence—but the Captain shows no other sign of exhaustion for some time—or a cut; the half-line has no obvious rhetorical significance, but since it is a complete half-line the words can continue smoothly without audible disruption of the verse, which resolves naturally on l. 23.)

33–5] OXFORD; *five lines ending* … assault | … and | *Banquoh* | … Eagles | … Lyon F; *three lines ending* … this | … Yes | … lion POPE (F had to run *Banquoh* over for lack of space, and then to split l. 35, so arranged as five lines. Editorial arrangements make technically regular lines, but suggest an irrelevant stress on the merely extra-metrical 'Yes'. Duncan may respond at once, or the Captain may pause for breath where he was expected to continue his story.)

37–8] F; *three lines ending* … cracks | So they | … foe BOSWELL, MUIR; *two lines ending* … they | … foe GLOBE (Putting 'So they' in a line alone has no rhetorical justification, transferring them to the previous line no merit; 'doubly' is probably an extra-metrical reduplication.)

41–2] F; *lines ending* … tell— | … help— ROWE (F had no need to realign; rearrangement exaggerates the pause after 'tell'.)

46–8] HANMER; *five lines ending* … eyes | … strange | … King | … Thane | … King F (The compositor had not enough space for l. 46; he got straight again after 'God save the King'.)

58–60] This edition; *four lines ending* … vs | … happinesse | … King | … composition F; *three lines ending* … happiness | … now | … composition STEEVENS 1778; *two lines ending* … now | … composition GLOBE; *three lines ending* … happiness | … Sweno | … composition FOAKES; *three lines ending* … happiness | … king | … composition HUNTER (I take it that Duncan's exclamation is extra-metrical, so that Ross continues his own line; the compositor seems to have set out to print 'That … composition' as one line but found he had not enough space.)

1.3.5] POPE; *two lines ending* ... mouncht. | ... quoth I F (Space.)

78] POPE; *two lines ending* ... greeting | ... you F (Space.)

81–2] CAPELL; *three lines ending* ... corporal | ... Winde | ... stay'd F (The compositor had not enough space to complete l. 81, so he fudged the next line as well.)

103–4] F; *as one line* SINGER

108–9] CAPELL; *three lines ending* ... liues | ... Robes | ... yet F (Space forced a division of l. 108, but order was restored unusually promptly.)

111–15] F; *four lines ending* ... combin'd | ... Rebel | ... both | ... not MALONE (Space in F would not have allowed Malone's arrangement, which produces rather awkward long lines; F moves just as well, although 111 and 115 are short and 113 is rather flat. It seems possible that F's arrangement derives from cuts.)

132–3] ROWE; *lines ending* ... good | ... successe F (There is no apparent space problem to explain F's arrangement, so it might possibly, as with ll. 141–3, spring from respecting the sentence structure more than the verse; or both could conceivably derive from MS.)

141–3] POPE; *lines ending* ... Man | ... surmise | ... not | ... rapt F

144] ROWE; *two lines ending* ... King | ... me F (Space.)

150–7] POPE; *twelve lines ending* ... fauour | ... forgotten | ... registred | ... Leafe | ... them | ... vpon | ... time | ... speake | ... other | ... gladly | ... enough | ... friends F (All this was clearly caused by justifying so as to stretch the scene to fill out the column.)

1.4.1–8] CAPELL (1–2), POPE (2–8); *nine lines ending* ... Cawdor | ... return'd | ... back | ... die | ... hee | ... Pardon | ... Repentance | ... him | ... dy'de F (It may be that MS had 'My Liege' extra-metrically in l. 3, for which the compositor had no space; at any rate it takes him until l. 8 to get straight.)

23–8] F; *five lines ending* ... part | ... Duties | ... Servants | ... thing | ... Honor POPE (There is no obvious need for F's arrangement; Pope may be right, though he characteristically prefers an awkward long line to short ones. I have restored F because there is an obvious need for ceremonial after ll. 22–3, and in the middle of l. 28.)

1.5.21–2] POPE; *three lines ending* ... winne | ... cryes | ... it F (Compositor had no space for the end of l. 19 and so carried it over.)

1.6.1–2] ROWE; *lines ending* ... seat | ... selfe F (Space did not require this: F follows grammar rather than verse.)

11–12] This edition; *lines ending* ... Hostesse | ... trouble F (This seems exactly like ll. 1–2 and so should be printed in the same way; it is possible that MS rather than compositor is responsible.)

18–21] POPE; *lines ending* ... broad | ... House | ... Dignities | ...

Ermites F (Again F follows grammar rather than verse, but here the reason may be justifying the foot of the column.)

2.1.5] ROWE; *two lines ending* ... Sword | ... Heauen F (Space.)

8–10] ROWE; *lines ending* ... sleepe | ... thoughts | ... repose F (There is no obvious reason for F's arrangement, and although the compositor is more than usually generous with spaces above and below the s.d. this comes at the head of a column, which makes justifying unlikely. F looks clumsy, but allows a clearer rhythm to 'That nature ... repose'; if, which I doubt, Banquo is seriously invoking the powers (cf. Lady Macbeth in 1.5) then a pause after 'sleep' would be appropriate. Rowe's rearrangement works perfectly well so I have, hesitantly, adopted it.)

14–15 pleasure, | And sent] F; pleasure, and | Sent CAMBRIDGE (Compositor A may have been dividing by syntax, but the alternative seems to me superior only to the eye, not to the ear.)

17–18] POPE; *lines ending* ... Hostesse | ... content F (No problem of space, so only grammar can explain the rearrangement.)

26–7] ROWE; *lines ending* ... consent | ... you F

2.2.1] them FI (thē) (A rare instance of Compositor A squeezing a line in by abbreviation rather than by carrying a word over.)

2–6] ROWE; *lines ending* ... fire | ... shriek'd | ... good-night | ... open | ... charge | ... Possets F; *lines ending* ... fire | ... shriek'd | ... night | ... open | ... Snores | ... Possets KNIGHT (F did not have room for l. 2 which may account for all that follows; Knight's variation seems to have no advantage.)

8–9] F (The lines can be, and often are, printed as one, but that obscures the fact that Lady Macbeth does not notice his entry until 'My husband?' Nevertheless, it does make a regular line of verse.)

14] ROWE; *two lines ending* ... don't | ... Husband F (F was not short of space; the arrangement was probably deliberate, perhaps in MS.)

15] ROWE; *two lines ending* ... deed | ... noyse F (A heavy pause would make sense, but F's division was necessitated by space.)

17–18] This edition; *all speeches as separate lines* F ('Ay' and 'Hark' are extra-metrical, often printed in separate lines.)

19–24] This edition; *seven lines ending* ... sight | ... sight | ... sleepe | ... other | ... Prayers | ... sleepe | ... together F; *six lines ending* ... sight | ... sight | ... Murther | ... them | ... them | ... together ROWE (Verse is not lost here, but it is not strictly sustained; Macbeth's train of thought seems to be continuous, whether or not he pauses at the end of l. 23, and l. 24 comes from Lady Macbeth's separate thoughts, not necessarily interrupting him. Decision is better left to actors than dictated by typography.)

30–2] POPE; *lines ending* ... Amen | ... throat | ... thought F (F would not have had space for l. 32, and so crowded the line before: this comes near the bottom of the column.)

64–5] POPE; *lines ending* ... white. *Knocke* | ... entry | ... Chamber F (F had not enough space for the s.d. *Knocke* within the line.)

71–3] POPE; *five lines ending* thoughts | ... deed, *Knocke* | ... selfe | ... knocking | ... could'st F; *three lines ending* ... deed | ... myself | ... could'st FOAKES (F had not enough space for l. 72 and the s.d., so divided the line but left '*Knocke*' against the first half; Foakes's arrangement is possible, but in effect it is only a question of placing the half line which is common rhythmically to both speeches.)

2.3.22–3] JOHNSON; *as verse, lines ending* ... Cock: | ... things F (F is presumably following MS, possibly misled by Macduff's verse, but a capital after a colon was possible anyhow.)

46–52] *Sometimes printed as verse in various unsatisfactory arrangements* (Lines 48 and 50 both have the metre of $1\frac{1}{2}$ lines of verse, used for the balanced courtesies. F uses capitals after colons which make these lines (like 22–3) look as if verse; the pitch is lowered towards prose before Lennox embarks on 55 ff. as a prelude to Macduff's return.)

55–7] ROWE; *lines ending* ... vnruly | ... downe | ... th'Ayre | ... Death F (Space did not strictly force this, but l. 55 would have been very full.)

60–2] HANMER; *four lines ending* ... time | ... Night | ... feverous | ... shake F (There is no obvious reason for this, so it is possible that the lineation of all of ll. 55–62 derives from MS; far more probable that it was a whim of Compositor A.)

71–2] This edition; *as three half-lines* F, OXFORD; *as two lines ending* ... life? | ... majesty? MUIR (Muir assumes that Macbeth and Lennox speak simultaneously, which is possible; but since half-lines frequently end speeches, my arrangement seems more likely—it is nearly simultaneous in any case; the usual arrangement gives Lennox's words a quite pointless prominence.)

83–5] This edition; *four lines ending* ... Businesse | ... parley | ... speake, speake | ... Lady F; *three lines ending* ... business | ... parley | ... lady THEOBALD (In any arrangement a pause is implied before Macduff speaks.)

88–9] THEOBALD; *lines ending* ... fell | ... murther'd | ... alas F

121–6] This edition; *nine lines ending* ... Lady | ... tongues | ... ours | ... here | ... hole | ... away | ... brew'd | ... Sorrow | ... Motion F (F's arrangement derives from space problems and is obviously awkward, so are all others; I assume this aside dialogue starts a new line and the rest follows, but l. 120 is long. It is possible that these lines were an afterthought, since they first articulate the suspicions that are previously only implicit in the dialogue, and F has similar problems with Malcolm and Donalbain in ll. 132–8.)

137–43] ROWE; *nine lines ending* . . . doe | . . . them | . . . Office | . . . easie | . . . England | . . . I | . . . safer | . . . Smiles | . . . bloody F (See note above.)

2.4.14] POPE; *two lines ending* . . . Horses | . . . certaine F (Space.)

19–20] POPE; *three lines ending* . . . so | . . . vpon't | . . . *Macduffe* F (Compositor B, helping out here and starting a fresh page, set 'to th' amazement' onwards; Compositor A presumably finished his page in mid-line for reasons of space, and so left a confusing legacy to B who does not usually tamper with verse in this way.)

32–3] This edition; *as three half-lines* F, OXFORD; *lines ending* . . . body | . . . Colmekill MUIR (The problem is merely typographical: there are three half-lines, not implying any pause; F printed each to the left as usual; so did Rowe and Pope, but because prefixes occupied part of the line, Ross's words look like a completion of Macduff's; most later editors printed it as such. I have preferred to place the odd one out at the end of a speech; see above, pp. 213–14.)

3.1.11–45] F (*substantially*) (The part-lines here are so numerous as to look odd, and cannot be reduced by rearrangement. The first two (13 and 18, set by Compositor B) may well be filled out with ceremonial welcoming; thereafter they correspond to abrupt changes of subject—if they signify more than that, it is probably pauses in the strained dialogue between Macbeth and Banquo. Cutting does not seem a likely explanation, despite the breakdown in lineation towards the end.)

34–5] POPE; *three lines ending* . . . Horse | . . . Night | . . . you F

41–3] THEOBALD (*subst.*); *lines ending* . . . societie | . . . welcome: | . . . alone: F (F associates 'To make . . . welcome' more closely with the previous words; in fact it goes equally well with either.)

45–50] HUNTER (*subst.*); *eight lines ending* . . . men | . . . pleasure | . . . Gate | . . . vs | . . . thus | . . . deepe | . . . that | . . . dares F (Almost as many solutions as there have been editors; it may be that the compositor went wrong in the previous dialogue and took several lines before he got right—but no solution is quite satisfactory. Irregularity, rather than cutting, is the likeliest explanation.)

71] F; *as two lines ending* . . . th'vtterance | . . . there (F may be making space at the end of his column.)

75–91] This edition; *as verse, lines ending* . . . then | . . . speeches | . . . past | . . . fortune | . . . selfe | . . . conference | . . . you | . . . crost | . . . them | . . . might | . . . craz'd | . . . *Banquo* | . . . vs | . . . so | . . . now | . . . meeting | . . . predominant | . . . goe | . . . man | . . . hand | . . . begger'd | . . . euer | . . . Liege F; *as verse, lines ending* . . . now | . . . Know | . . . you | . . . been | . . . you | . . . Instruments | . . . might | . . . craz'd | . . . us | . . . now | . . . finde | . . . nature | . . . Gospell'd | . . . Issue | . . . Grave | . . . Liege ROWE (F is obviously altogether wrong, but Rowe's rearrangement is clumsy, often flat, and involves two awkwardly long lines. Elsewhere I have followed similar reconstructions, sometimes reluctantly, but here I think the passage is effectively prose, in strong contrast to the verse before it (despite a number of iambic feet); Macbeth's rhetoric leads to a blank verse cadence at the end of his speech (ll. 82–3) but is returned to prose by the Murderer's flat response, then finally explodes into verse again at l. 92.)

114–15] ROWE; *three lines ending* ... on't | ... Enemie | ... Lord F (F prints Macbeth's words as one line.)

128] POPE; *two lines ending* ... you | ... most F (The line is long and F could not have printed it as one, but two lines implying a pause is possible.)

3.2.17–18] F; *as one line* POPE

24–5] F; *as one line* ROWE (Many editors have expressed uncertainty about 17–18 which makes a very long single line (a hexameter with a feminine ending) and could not have been so printed in F, whereas 24–5 arouses no comment though it creates an alexandrine for which F had plenty of space. There is no other sign of justifying here and the division does correspond to a strong break in thought, so a pause may well be implied. The irregularities present no problem in delivery, and seem particularly sensitive in this crucial scene.)

35–7] POPE; *four lines ending* ... laue | ... streames | ... Hearts | ... are F; *four lines ending* ... we | ... streams | ... hearts | ... are STEEVENS (There is no obvious reason why F rearranged here, but though Steevens's arrangement displaces the fewest words, it is unsatisfactory; Pope's works straightforwardly enough.)

46–7] ROWE; *lines ending* ... Peale | ... note F (Space probably explains F's arrangement though it was not absolutely necessary.)

3.3.9–10] POPE; *lines ending* ... hee | ... expectation F (In the rest of the scene, though many lines suggest verse rhythm it is not consistent, which is not surprising—and only flashes of it will be heard amid the violent action in the theatre. Arrangement varies greatly with editors and is not significant. F spreads it out as far as possible to fill the column which was the last in Compositor A's half of the play.)

16–18] This edition; *lines ending* ... Trecherie | ... flye, flye | ... Slaue F; *two lines ending* ... fly, fly | ... slave HANMER

21–2] CAPELL; *lines ending* ... Affaire | ... done F

3.4–5.7: *Compositor B*

3.4.1–8] This edition; *as verse, lines ending* ... downe | ... welcome | ... Maiesty | ... Society | ... Host | ... time | ... welcome | ... Friends | ...

welcome F; ... first | ... Majesty *rest as* F CAPELL (*conj.* Johnson) (F's arrangement is hopeless and editors' no improvement. The confusion looks superficially like the work of Compositor A, but Hinman is in no doubt that Compositor B took over here. Werstine (see n. 1, p. 214 above) shows that B was prone to print prose as verse and my assumption is that he did so here; verse is already anticipated in iambic syllables and from l. 9 onwards, though still fairly prosaic, the verse is in fact regular (see p. 216 above). Subsequent confusions in ll. 12 ff. are quite characteristic of B (see below) and look distinctive only because of their proximity to each other and to the opening lines.)

12–16] ROWE (*subst.*); *lines ending* ... face | ... then | ... within | ... dispatch'd | ... him | ... Cut-throats F

19–21] F; *as two lines ending* ... 'scaped | ... perfect POPE (12–16 follows the usual F practice of conflating half-lines in short speeches and the usual editorial arrangement is convincing. But for 19–21 it is not; F's arrangement is so eccentric that it is most easily explained either as justifying, which is unlikely in the middle of a column, or as indicating very long pauses: the Murderer's reluctance to admit failure, and Macbeth's shock on hearing of it.)

48] CAPELL; *two lines ending* ... Lord | ... Highnesse F (Space.)

69] CAPELL; *two lines ending* ... there | ... you F

110–11] ROWE; *lines ending* ... mirth | ... disorder | ... be F (There was not space for all of 110 and the second half was conflated with the first half of 111 in a single line.)

123] ROWE; *two lines ending* ... say | ... have Blood F (Space.)

3.6.1] ROWE; *two lines ending* ... Speeches | ... Thoughts F

4.1.85] ROWE; *as two lines* Macbeth, Macbeth, Macbeth | Beware *Macduffe* F (Space.)

93] ROWE; *two lines ending* ... resolute | ... scorne F (Space.)

100–1] ROWE; *lines ending* ... Thunder | ... King F (F puts 'What is this' normally to the beginning of the line after the s.d., and then fills out with the whole of 101.)

148] ROWE; *two lines ending* ... Gone | ... houre F (Space.)

4.2.27] ROWE; *two lines ending* ... is | ... Father-lesse F (He had plenty of space, so presumably the compositor was simply filling out to end the column neatly with Ross's exit.)

31–53] This edition; *as verse, lines ending* ... dead | ... liue | ... Mother | ... Flyes | ... they | ... Bird | ... Lime | ... Gin | ... Mother | ... for | ... saying | ... dead | ... Father | ... Husband | ... Market | ... againe | ... wit | ... thee | ... Mother | ... was | ... Traitor | ... lyes | ... so | ... Traitor | ... hang'd F; *as verse, various rearrangements* editors (The lines can make a fair show of verse, however unconvincingly, and there is no consensus on how they should go. F sets out 36–7 as three lines ending '. . . Bird | . . . Lime | . . . Gin', which may suggest a song but certainly not blank verse. In any case there is a sharp contrast of tone after Ross's exit, and most editors accept prose after l. 38 although F continues as if verse up to at least l. 53 and does not make prose clear until l. 58. It may well be that the compositor had set up lines exactly as in MS, but using initial capitals.)

60–1] POPE; *as verse, lines ending* . . . Monkie | . . . Father F (B evidently misinterpreted MS here, since he had space for more words in his first line.)

81–2] F; *as two lines, ending* . . . faces | . . . husband ROWE (Rowe's long line has no obvious value here; the entry breaks off Lady Macduff's speech, and she speaks again after a pause.)

4.3.25] ROWE; *two lines ending* . . . there | . . . doubts F (Space.)

102–3 Fit . . . miserable] POPE; *as one line, abbreviating to* 'Natiō' F (The first instance of Compositor B compressing the type; he does it twice later.)

139–40] MUIR; *three lines ending* . . . reconcile | . . . forth | . . . you F; *two lines ending* . . . reconcile | . . . you ROWE (F probably intended Malcolm's words as one line, carried over for lack of space.)

173–4] THEOBALD; *three lines ending* . . . sicken | . . . true | . . . griefe F (F prints Macduff's two part-lines as one.)

211–13] CAPELL; *five lines ending* . . . too | . . . found | . . . too | . . . said | . . . comforted F (F saves space by printing Ross's and Macduff's words as single lines, and very likely MS had them so too.)

215–17] F; *lines ending* . . . children | . . . say all | . . . chickens HANMER

218–20] F; *lines ending* . . . swoop | . . . so MUIR (Following F more closely than is usual gives all the pauses to Macduff, which seems appropriate—i.e. before 216, during 217, and very likely before or after 220, though 'Dispute it like a man' may float between 219 and 220.)

5.1.44–5] POPE; *two lines ending* . . . too | . . . not F (B presumably misinterpreted MS as verse: he capitalized 'You'.)

5.6.1] ROWE; *two lines ending* . . . enough | . . . downe F (Space.)

5.7.84] ROWE; *two lines ending* . . . art | . . . stands F (Space.)

MUSICAL ADDITIONS

1. *Song: 3.5.36–68, 'Come away, Hecate'*
The words of the songs in 3.5 and 4.1 have not hitherto been fully edited; indeed, the first scholarly edition in which they have been given as part of the text is the Oxford *Complete Works* (1986). Wells and Taylor used the version found in Crane's manuscript of Middleton's play, *The Witch*, collated with Davenant's version; but there are other early versions to be considered for the first song, 'Come away, Hecate'. Two manuscripts of this survive in collections of lute songs, but tentatively dated *c*.1630; their words are generally similar to those in *The Witch*, and to each other, and they give the same melody, but differ substantially in their bass lines for accompaniment. One is in the Fitzwilliam Museum at Cambridge, MU. MS 782, known as the Bull Manuscript; the other is in New York Public Library, Drexel MS 4175. The latter has been edited, with keyboard accompaniment, by Ian Spink in *The English Lute Songs*, 2nd series, vol. xvii: *Robert Johnson* (1974), 58–61. He states that it 'seems preferable' to Bull, but does not give reasons; its only substantial difference is in the accompaniment, otherwise it is marginally more careless both with words and music (particularly in marking naturals), but more interestingly seems towards the end to have had difficulty in deciphering its source: its first shot at l. 62 is partially deleted, 'and' becomes 'our', and it then brackets alternative epithets, the upper barely legible but probably 'cristell', the lower 'mistris'. Bull here reads 'and misty', *Witch* 'our Mistris'.

Neither manuscript refers the song to a play, nor to a composer, and since they offer it for a solo singer, he is left to mimic dialogue as he pleases, without indications for different voices. In both, this song occurs in a group where others are known to be by Robert Johnson, who composed for the King's Men from *c*.1609 to *c*.1615 (including music for *The Winter's Tale* and *The Tempest*). Another song from *The Witch*, not part of *Macbeth*, survives in a setting by Thomas Wilson, who is known to have worked for the company from about 1615. Wilson was only 20 in 1615, so he is hardly likely to have been involved earlier. Spink suggests that Johnson and Wilson collaborated on *The Witch* in 1615 because of Wilson's inexperience, which is possible; but it is equally possible that Johnson set 'Come away, Hecate' for *Macbeth* at an earlier date, and that Wilson provided the additional music for *The Witch* later.

'Come away, Hecate' first appeared in print in a quarto edition of *Macbeth* in 1673, which also gave additional songs for the witches after

2.2 and 2.3, but did not give words for 'Black Spirits' in 4.1; in all other respects it was a simple reprint of the first Folio. The title-page does not mention Shakespeare, nor Davenant, but does state that it was performed at the Duke's Theatre; it is followed by a cast list which is identical with that in another quarto of 1674, whose text is generally identical with the Yale manuscript of Davenant's version, though he is still not mentioned on the title-page which claims to give 'all the alterations, amendments, additions, and new Songs'. No doubt 1673 was produced to cash in on the celebrated revival of Davenant's version at the new Dorset Garden Theatre in 1672, and 1674 was intended to displace it. The songs in Act 2 appear in Davenant's version as part of an entirely new scene at the end of the act for the Macduffs' encounter with the witches, and in it they are separated by only five lines of dialogue. It seems that the printer of 1673, Cademan, who obviously did not have a full text of Davenant's version, invented his own locations for the songs, for it is absurd to put the first one in 2.2 when it refers to the murder as committed twelve hours before, 'Long ago, long ago'. The only substantial difference in the texts of these songs is that the last quatrain of the second song is not in 1673; this would seem a mere accident of piracy, were it not that it is singularly weak, and the song seems to me complete without it.

The versions of 'Come away, Hecate' in 1673 and 1674 are very similar verbally (not identical), but the allocation to voices is strikingly different. 1673 uses only numbers to distinguish them, and '2' must be Hecate; the questions in 3.5.41 and 42 are assigned to voice '1', and the responses to '3' and '4'. 1674 uses 'Hec.' for Hecate, and numbers three spirit voices; the questions are assigned to Hecate and the responses to voices '2' and '3'. Thereafter 1673 has no prefix for l. 46 which must be Hecate, as 1674 and Yale agree; all three give l. 48 to voice '1', but differ again in l. 52 which 1674 gives to voice '2', while both 1673 and Yale leave it with voice '1', and both prefix l. 53 with '2'—but in 1673 this should mean Hecate, in Yale it should not. None of the Restoration texts prefixes l. 55, which must be Hecate's, and they all give l. 58 to voice '3'; by this time 1673 seems to be using the same numbering system as 1674 and Yale: its voice '4' never utters after l. 42, but it is alone in marking the last four lines as 'Chorus'. Clearly there is confusion, and the identical numbering in the second half of the song probably does not mean different voices for 1673, though it should; on the other hand, the differences in the first part do appear to be deliberate. If Locke's music was new in 1672, 1673 might be supposed to show traces of an earlier setting, but I do not find conclusive evidence of this. The songs in 1673 appear to be pirated from Davenant's version and not, which would be more interesting, a relic of earlier expansion of the musical additions to the play.

The question remains where Davenant derived his text of the original songs in 3.5 and 4.1 from. They were certainly part of *Macbeth* before 1623, and though the Folio printers had no texts for them, the theatre presumably had; Davenant could have seen them when he worked for the King's Men in the 1630s, and in any case he presumably inherited the prompt-book when *Macbeth* was allocated to his company in late 1660. There is no reasonable doubt that he took the songs as part of Shakespeare's text; his texts differ only trivially from those of the 1620s, and almost entirely in ways that suggest accidents in copying rather than deliberate revision.

For the most part, the minor variations between the six texts are casual, and I can find no pattern to establish a chain or tree relating them. Lines 41 and 42 could relate Davenant to Bull, Drexel to *Witch*; l. 48 could relate 1673 to Bull, Davenant to Drexel and *Witch*; in l. 60, 'dance' occurs in Davenant and *Witch*, but not in either Bull or Drexel; and so on. The one interesting variant is in l. 62, where Drexel shows alternative readings, the less obvious one identical with *Witch*: Drexel's 'our' replaces a deleted word, probably 'and'; the bracketed alternatives 'cristell'/'mistris' could both be misreadings of 'misty' as in Bull, if it was spelt 'mistie' in Drexel's copy ('cr' and 'm' are very similar in these hands). Drexel seems to be struggling with a difficult manuscript, but one of his offers coincides with *Witch* ('our mistris'). 'Our mistress' ' offers the ingenious conceit of the seas as the moon-goddess's fountains, but the reading presents two problems: 1. that such wit is alien to the bland language of the rest of the song; 2. that whereas Hecate in *Macbeth* is certainly the goddess, in *The Witch* she is no more than a superior bawdy witch. Bull's 'misty fountains' is altogether more likely in this context, though it offends principle to attribute the wittier reading to a copyist. 'Crystal' is probably an irrelevance due to Drexel's efforts to decipher his copy, and offering a standard adjective for 'fountains'. 'and misty fountains' recurs in all three Restoration texts, but linked in them to a corruption of 'seas' into 'hills'—presumably a copyist's error at some stage, suggested by 'rocks and mountains' in the line above, producing a mere redundancy. It is difficult to see how Crane could have derived his reading from Drexel even if they had a common source, since elsewhere Drexel agrees with Bull against *Witch*. In short, speculation on such limited evidence cannot lead to safe conclusions.

I have treated Bull as my copy-text for two reasons: first, that it is marginally superior to Drexel, and both are closer to the music than is *Witch*; second, that it is slightly closer to the Restoration texts, and so might be associated with *Macbeth* rather than with *The Witch*. Variants in the other five texts are all shown in the collation.

2. *Song:* 4.1.44–59, '*Black Spirits*'

The second song—'Black Spirits' in 4.1—is less complicated because the only early manuscript is *The Witch*, and 1673 gives no text. We have therefore only *Witch*, Yale, and 1674. They are, again, very similar and it is logical to suppose that, as with 'Come away, Hecate', Davenant made few, if any, deliberate alterations. But there is here less guidance as to whether variants should be attributed to corruption in his copy, or in Crane's. Line 56 is given in *The Witch* to Firestone, but this time it is not an interpolation. Crane's 'Liand' in l. 48 is clearly an error for 'Liard' which is the name of a spirit in *Scot*, and in the pamphlet he was using; in Davenant it has become an adjective 'liar', a corruption that may have influenced his reduction of the spirits named from six to three. More puzzling is l. 52: Oxford prefers 'a grain', though noting that Crane's 'againe' makes sense, but rejects 'lizard's brain' because it could be a corruption of Crane's 'Libbard's bane'; but it is not clear which is in error, and it is odd that *The Witch* refers in dialogue just before this song to lizard's brain. Both forms are entirely possible: Leopard's bane is still used of a common woodland plant (also known as Herb Paris) which has a single black berry nestling in dark green leaves, looking sinister though in fact it is not poisonous. Lizards, like snakes, were commonly associated with witchcraft. The fabulous basilisk resembled a lizard, and dispensed death through its eyes, the emanation of its brain; Paracelsus alluded to its generation from menstrual blood, supposed to be 'the greatest impurity of women' (H. Silberer: *Hidden Symbolism of Alchemy and the Occult Arts* (New York, 1917), 139). Frazer gives an account of a modern gypsy rite in southern Europe, for an annual expulsion of evil spirits, which included dried lizard in a brew largely of herbs. Neither substance is in *Scot*, and it seems to me likely that 'lizard's brain' in Middleton's text derives, like the names of his witches, from the song. On the whole, then, I think the error probably Crane's, and it is more consistent to follow Davenant for both readings. In l. 54, however, Davenant's 'the charm grow madder' looks like a feeble substitution for Crane's 'younker', doubtless because of the oddity of applying 'younker' to Macbeth. The word is entirely appropriate to *The Witch*, where the object of the song is to seduce a youth; if it ever stood in *Macbeth*, its sense must have been ironic. I have retained it because Davenant's line is unconvincing, and it is a likely part of the children's game (see below). In l. 15 again Davenant's line looks like a feeble rationalization of *Witch*'s; but here it must be said that Crane's reading is obscure, all the more so since he has a full stop after 'all'.

The only other question is who sings. Crane gives no prefixes until l. 50 (for 'Put in that'), and from then on he distributes lines between Hecate and two witches, exactly as Davenant does (except for l. 56). This is plausible in *The Witch*, but in *Macbeth* it comes oddly after the direction 'Enter Hecate and the other three Witches'. The number *may* be an error (see note on l. 38.1), but if it is correct that three enter, why do only two sing? The third cannot be a silent ballerina, since they evidently dance all together round the cauldron. In *Macbeth*, Hecate commands her witches to sing and dance, and might not be supposed to join in herself. This fits her detached silence throughout the scene until its closure; otherwise she could logically supply the commands in ll. 51, 52, and 55 which Crane attributed to her. Yale gives her these and also ll. 48 and 56 which is less plausible (1674 leaves ll. 47–8 with the First Witch). In Davenant's version, Hecate takes exclusive charge of the predictions which follow, so her participation here is quite likely; there is no trace of such a change in F. Davenant's Hecate can thus be given l. 56, but it is inconsistent to deprive her of the command in l. 55.

I have arranged 'Black Spirits' simply for three voices, presumably the 'three other witches'. They would doubtless be the same singers as for 3.5, quite possibly children 'Like elves and fairies in a ring' (4.1.42).

'Come away, Hecate' was probably written specially for whichever play first used it; the origin of 'Black Spirits' is less certain. Its immediate source appears to be in *Discovery of Witchcraft*, but Scot's passage has the peculiarity that it is written in a jingle which Scot did not find in his own source: 'he-spirits and she-spirit's, Tittie and Tiffin, Suckin and Pidgin, Liard and Robin, etc.: his white spirits and black spirits, gray spirits and red spirits, devil toad and devil lamb, devil's cat and devil's dam' (ed. Brinsley Nicholson (1886), p. 542). Nicholson, in his monumental edition of Scot, speculated that this passage might have been influenced by a ballad because verse is entirely uncharacteristic of the author. He certainly did not find it in his pamphlet source *A true and just Record, of the Information, Examination and Confessions of all the Witches, taken at S.Oses [St Osyth] in the county of Essex* (1582). The material of Scot's passage is tabulated (with more detail) in a fold-out sheet at the end:

> Tettey a he like a gray cat, Jack a he like a black cat;
> Pygin a she like a black frog, and Tyffyn a she like a white lamb;
> Suckytt a he like a black frog; and Lyard red like a lion or hare;
> two spirits like toads, their names Tom & Robin;
> iiii spirits, viz. their names Robin, Jack, Will, Puppet
> ... wherof two were hes and two shes, like unto black cats.

Scot's antithetic structures might have been suggested by this, but not his doggerel verse. Crane heads this song in *The Witch* 'A Charm Song: about a Vessel' which is redundant as a direction in the play, though an accurate description: it reads like a singing game such as children play, leading to a girl claiming her own choice of boy. This peculiarity had

15, 16. Robert Johnson's setting of 'Come away, Hecate' in 3.5, from the Bull Manuscript.

struck me before I became aware of Nicholson's comments, and led me to wonder whether it had an independent source as a well-known rhyme. That would explain the disconcerting 'younker' in l. 54, and perhaps the definite article of 'the red-haired wench' if that once referred to the Magdalen (see note on this line); otherwise Davenant's modifications must seem to be of *The Witch* itself, which is hardly probable.

The Bull manuscript of 'Come away, Hecate' is reproduced here in Figs. 15 and 16. A transcription of the Drexel manuscript is in J. P. Cutts, 'Robert Johnson: King's Musician in His Majesty's Entertainment', *Music and Letters*, 36 (1955), 111–16; Spink's edition is based on Drexel.

Repeats marked in the music appear to have no significance; those marked for the words are clearly required by the music; the values of notes as marked vary, but this does not seem to be confusing. After the first line, the manuscripts diverge substantially in the bass, but for the melody differences are slight, mostly minor preferences of arrangement; in the second bar of the fourth line Drexel is probably correct in reading 'F', since Bull's 'E' is incompatible with 'F' in the bass. I have not attempted a full collation of the score; Spink notes variants when his edition departs from Drexel.

3. *Dance: 4.1.147.1*

It has often been suggested that music and choreography for the Witches' dance before they finally disappear might have been taken over from the antimasque of Jonson's 'Masque of Queens', performed at Court in 1609. J. P. Cutts in 'Jacobean Masque and Stage Music', *Music and Letters*, 35 (1954), 191–3, supports this, asserting that Johnson's music for 'Oberon' was used in *The Winter's Tale*, and that for the madman's antimasque in 'The Lord's Masque' in *The Duchess of Malfi*. He goes on to discuss music for the first and second witches' dances which form items 51 and 52 in British Library Additional MS 10444. The first of these was printed in Dowland's *Varietie of Lute Lessons* in 1610 as 'The Witches' Dance in the Queen's Masque', but there are difficulties in identifying it as Johnson's music: Cutts finds it curiously lacking in unusual rhythmic movement, whereas the second dance is much more vigorous and like Johnson's acknowledged dances for satyrs, etc., in the manuscript. The matter seems to be uncertain, and I have not attempted to reproduce the music for either dance, but Ben Jonson's account of the witches' performance in the 'Masque of Queens' may well give an idea of the choreography in *Macbeth*:

> At which, with a strange and sudden music, they fell into a magical dance, full of preposterous change and gesticulation; but most

applying to their property [i.e. most applicable to their proper nature] who, at their meetings, do all things contrary to the custom of men, dancing back to back, hip to hip, their hands joined, and making their circles backward, to the left hand, with strange fantastic motions of their heads and bodies. All which were excellently imitated by the maker of the dance, Mr. Jerome Herne, whose right it is here to be named.

In the heat of their dance, on the sudden was heard a sound of loud music, as if many instruments had given one blast: with which not only the hags themselves but their Hell, into which they ran, quite vanished—and the whole face of the scene altered, scarce suffering the memory of any such thing ... (*Works*, vii, ed. C. H. Herford and P. and E. Simpson (Oxford, 1941), 301.)

APPENDIX C

MACBETH AT THE GLOBE, 1611

SIMON FORMAN, well-known as dubious astrologer, dabbler in magic and witchcraft, and as successful (even heroic) doctor to poor as well as rich, began a volume of notes on his playgoing in April and May 1611, which he titled 'The Book of Plays and Notes thereof by Forman for Common Policy [i.e. public morals]'. He only recorded, scrappily, four plays: *Macbeth*, *Cymbeline*, *The Winter's Tale*, and a *Richard II* which was certainly not Shakespeare's. The last two he dated 1611, *Cymbeline* is undated, but *Macbeth* he claims to have seen in 1610, on Saturday 20 April. 20 April was not a Saturday in 1610, but it was in 1611, and no doubt 1610 was a slip. The year was still commonly dated in the old style from Lady's Day, 25 March, and slips must have been common in April (as nowadays they are in January). Wilson remarked that performances rarely, if ever, took place before May; but they depended more on weather than calendar, and good weather still occasionally happens in April.

A number of oddities led to the suspicion that the document was a forgery. It is in Oxford, Ashmolean MS 208, but it was first brought to light in the nineteenth century by J. P. Collier, in a collection of materials several of which were certainly forged. Authenticity was finally claimed by J. Dover Wilson in *The Review of English Studies*, 23 (1947), 193–200, of which the last and most conclusive part is by R. W. Hunt. The handwriting was genuine, and a copy of these pages had, as Collier claimed, been supplied to him by H. W. Black (who was engaged in cataloguing the Ashmolean manuscripts) in 1832.

Unfortunately, although it is genuine, internal evidence still suggests that it is not a reliable account of performance. Forman's memory was both confused and erratic, and there are strong reasons for thinking he had looked at Holinshed as well as at the stage. It is highly unlikely that he saw Macbeth and Banquo on horseback, and if he had they would not have been 'riding through a wood'. Shirley (pp. 168–89) discusses this fully, and concludes that the most that was likely was a noise of horses' hooves off-stage before Macbeth and Banquo enter (almost no directions call for actual horses on stage). But they *are* mounted, beside a large tree, in an illustration to the first edition of Holinshed (Bullough, p. 494); and in a nearby passage of his text, Holinshed describes the three women they encounter as, amongst other things, 'nymphs or fairies' (Bullough, p. 495). So does Forman, though the words are not in the play, not likely ever to have been, and not likely to have occurred to anyone as a description of the Weïrd Sisters as played by male actors in 1611.

It follows that we cannot accept this, as we would like to, as an (almost unique) account of a performance at the Globe. No doubt Forman did not see the Macbeths trying in vain to wash the blood off their hands, but too much has been made of that, for his comment is a fair imaginative response (especially after he had seen Lady Macbeth sleep-walking). But if his account is not reliable in what it does say, it is certainly not evidence when it passes over major incidents without mention. Hecate may or may not have been seen at the Globe in 1611, but the apparitions must have been; neither is mentioned by Forman at all. In fact, he is fairly detailed as far as 3.4, the appearance of Banquo's ghost; but after that his attention was extremely perfunctory, and it is odd that he found nothing for common policy in the play's conclusion—though he does backtrack to mention the sleep-walking.

For these reasons, despite its obvious interest, I have not included this passage in the Introduction, but give it here in modern spelling (Forman's own was very eccentric, and certainly not derived from Holinshed). The manuscript is reproduced and transcribed in S. Schoenbaum, *Shakespeare: Records and Images* (1981), 7–20.

The Book of Plays and Notes thereof per Forman for Common Policy

In *Macbeth* at the Globe 1610, the 20 of April, Saturday [indicated by the astrological sign for Saturn], there was to be observed first how Macbeth and Banquo, two noblemen of Scotland, riding through a wood, there stood before them three women, fairies or nymphs, and saluted Macbeth, saying three times unto him: 'Hail Macbeth, king of Codon [i.e. Thane of Cawdor], for thou shalt be a king, but shall beget no kings, etc.' Then said Banquo: 'What, all to Macbeth, and nothing to me?' 'Yes' said the nymphs, 'hail to thee, Banquo, thou shalt beget kings, yet be no king.'

And so they departed and came to the court of Scotland, to Duncan, King of Scots, and it was in the days of Edward the Confessor. And Duncan bad them both kindly welcome; and made Macbeth forthwith Prince of Northumberland [i.e. Cumberland; Londoners still often confuse the two], and sent him home to his own castle; and appointed Macbeth to provide for him, for he would sup with him the next day at night, and did so. And Macbeth contrived to kill Duncan, and through the persuasion of his wife did that night murder the king in his own castle, being his guest. And there were many prodigies seen that night, and the day before. And when Macbeth had murdered the king, the blood on his hands could not be washed off by any means, nor from his wife's hands, which handled the bloody daggers in hiding them; by which means they became both much amazed and affronted.

The murder being known, Duncan's two sons fled, the one to England, the [other to] Wales, to save themselves. They being fled, they were supposed guilty of the murder of their father, which was nothing so. Then was Macbeth crowned king; and then he, for fear of Banquo his old companion, that he should beget kings but be no king himself, he contrived the death of Banquo, and caused him to be murdered on the way as he rode. The next night, being at supper with his noblemen whom he had bid to a feast, to the which also Banquo should have come, he began to speak of noble Banquo, and to wish that he were there. And as he thus did, standing up to drink a carouse to him, the ghost of Banquo came and sat down in his chair behind him; and he, turning about to sit down again, saw the ghost of Banquo, which fronted [i.e. affronted] him so, that he fell into a great passion of fear and fury, uttering many words about his murder by which, when they heard that Banquo was murdered, they suspected Macbeth.

Then Macduff fled to England to the king's son. And so they raised an army and came into Scotland, and at Dunsinan ['dunston Anyse'] overthrew Macbeth. In the meantime, while Macduff was in England, Macbeth slew Macduff's wife and children; and after, in the battle, Macduff slew Macbeth.

Observe also how Macbeth's queen did rise in the night in her sleep, and walk, and talked and confessed all, and the doctor noted her words.

INDEX

In this index, the page numbers indicate subjects and people discussed in the Introduction; the act/scene references indicate words, phrases, and people discussed or glossed in the Commentary; and 'App.' indicates the Appendices (detailed references are not given). Biblical and proverbial allusions are grouped together. An asterisk indicates that the note supplements information given in *OED*.

Abbott, E. A., 1.2.10–12
absolute, 3.6.41
Acheron, pit of, 3.5.15
actual, 5.1.12
addition, 1.3.106; 3.1.100
addressed, 2.2.23
adhere, 1.7.52
admired, 3.4.111
advantage, 5.4.11
affection, 4.3.77
affeered, 4.3.34
after, 5.1.14
agitation, 5.1.11
aid, upon his, 3.6.30
air, 3.5.20
air-drawn dagger, 3.4.62
alarm/alarum/alarumed, 1.2.0; 2.1.54; 5.2.4
Aleppo, p. 62
all-hail, 1.5.54
All's Well That Ends Well, p. 49; 2.3.18
allthing, 3.1.13
amazed, 5.1.76
angel/angels, p. 9; 5.7.44
*annoyance, 5.1.74
anon, 1.1.10
an't, 3.6.19
antic round, 4.1.145
anticipat'st, 4.1.159
Antony and Cleopatra, pp. 12, 20, 59, 63–4; 3.1.56, 5.7.31
appals, 2.2.57
apparitions, pp. 5, 74
approve, 1.6.4
armed, 3.4.102
armed head, 4.1.82.2
arms, 5.7.18–19
aroynt, 1.3.6
As You Like It, 3.4.106
assay, 4.3.143
auger-hole, 2.3.124
augures, 3.4.125
authorized, 3.4.66

baboon, 4.1.37
baby of a girl, 3.4.107
badged, 2.3.104
bane, 5.3.58
bank, 1.7.6
banquet, 3.4.0
Banquo: and Weird Sisters, pp. 2–3, 4; descendants of, pp. 68, 73–4; ghost of, p. 4; 3.4.37.1–2
Barnes, Barnabe, *The Devil's Charter*, p. 63
baroque art, pp. 24–31, 34
baroque drama, pp. 22–34
bat, 3.2.43
bath, 2.2.37
battle, 5.6.4
beards, 1.3.46
bear-like, 5.7.2
Beaumont, Francis, *The Knight of the Burning Pestle*, p. 63
Beckett, Samuel, *Waiting for Godot*, p. 56
become, 1.7.46
bed, 1.6.8
Beelzebub, p. 80; 2.3.3–4
beguile the time, 1.5.62
behold, 5.3.20
beldams, 3.5.2
Belial, p. 80
Bellenden, John, p. 67
bellman, fatal, 2.2.3
Bellona's bridegroom, 1.2.54
bend up, 1.7.80
benison, 2.4.40
Bernard, Saint, His Meditations, 2.2.59–60
Bernini, Giovanni, pp. 28–31, 32, 34; *Cornaro Chapel*, pp. 28–31; *Ecstasy of St Teresa*, pp. 28–31, 34
betimes, 3.4.134; 4.3.162
better part of man, 5.7.48
Betterton, Thomas, pp. 40, 41

Biblical allusions, p. 80; 2.3.3–4;
  2.3.66; 2.3.132; 3.1.16; 4.2.33;
  5.5.23–4; 5.5.24–5
bill, 3.1.100
Birnam Wood, p. 5; 4.1.108
birthdom, 4.3.4
birth-strangled, 4.1.30
Blackfriars Theatre, pp. 1, 34–5, 63,
  64, 66; 3.5.47.1
blanket, 1.5.52
blaspheme/blaspheming, 4.1.26;
  4.3.108
blast, p. 9; 1.7.22
bleeding, 5.2.4
blind-worm, 4.1.16
blood-baltered, 4.1.138
bodements, 4.1.111
Boethius, Hector, *Chronicle of
  Scotland*, pp. 67, 69, 71
bond, 3.2.52
*borne in hand, 3.1.80
bosom, 2.1.29
bosom interest, 1.2.65
botches, 3.1.134
Bothwell, James Hepburn, Earl of,
  p. 78
Bradley, A. C., pp. 10, 22–3;
  1.3.122
braggart, 4.3.231
Brathwait, Richard, *Survey of
  History*, 1.6.4
break, 1.7.48
breath, 4.1.114–15; 5.3.27
breeched, 2.3.116–18
brevity of *Macbeth*, pp. 55–6
brewed, 2.3.125
brinded, 4.1.1
broad, 3.6.21
broil, 1.2.6
Bromham, A. A., p. 64
Brooks, Cleanth, pp. 8–9, 23
Brown, John Russell, p. 49
bruited, 5.7.23
Buchanan, George, *Rerum Scoticarum
  Historia*, p. 67
Buck, Sir George, p. 72
Bullough, Geoffrey, pp. 66–7, 76
Burbage, Richard, p. 41

cabined, 3.4.24
*called, 1.3.39
candles, 2.1.6
Captain, 1.2.3

Caravaggio, Michelangelo, pp. 24–6,
  27, 32, 34; *Conversion of Saul*,
  pp. 24–6; *Martyrdom of St
  Matthew*, pp. 26, 27
card, 1.3.17
Carracci, Annibale, p. 24
casing, 3.4.23
cast the water, 5.3.49
catalogue, 3.1.92
cause/causes, 3.1.33–4; 5.2.3
celebrates, 2.1.52
censorship, political, pp. 61, 71, 75,
  76
censures, just, 5.4.14
ceremony, 3.4.33–7
chalice, p. 8; 1.7.11
chamberlains, 1.7.64
chambers, 5.4.2
champion, 3.1.71
Chapman, George, *Bussy d'Ambois*,
  p. 61; 4.1.56; *Eastward Ho!*, p. 61
charge, 4.3.20
Charles I, King, p. 34
charnel houses, 3.4.71
chawdron, 4.1.33
cheer, 5.3.21
cherubim, 1.7.22
Chester miracle plays, p. 80
choke their art, 1.2.9
choppy, 1.3.44
chops, 1.2.22
choughs 3.4.126
Christian, King of Denmark, p. 72
chuck, 3.2.48
cistern, 4.3.63
Clarendon edition (W. A. Wright),
  p. 57
classical sources for *Macbeth*,
  pp. 76–8
clear/clearness, 1.5.70; 1.7.18;
  3.1.133
clept, 3.1.94
cling, 5.5.40
clogs, 3.6.44
close, 3.2.15; 3.5.7; 5.1.19
closed, 3.1.99
closet, 5.1.6
Clytemnestra, p. 77
coign of vantage, 1.6.7
Coleridge, S. T., p. 23; 1.5.52
coll, 3.5.49
Colmekill, 2.4.33
colours of their trade, 2.3.116–18

combustion, 2.3.59
come in time, 2.3.5
command upon me, 3.1.16
commends/commendations, 1.4.56;
 1.7.11
commission, in, 1.4.2
common enemy of man, 3.1.68
compassed, 5.7.86
composition, 1.2.60
compunctious, 1.5.44
conceive, 2.3.66
consequence, p. 7
consequences, mortal, 5.3.5
consort, 2.3.137
constancy, 2.2.67–8
continent impediments, 4.3.64
convey, 4.3.71
convince/convinces, 1.7.65; 4.3.142
copy, 3.2.41
Coriolanus, p. 20; 2.3.4–5; 4.2.17
corporal, 1.3.81
corporal agent, 1.7.81
count, in, 1.6.27
countenance, 2.3.82
country's honour, 3.4.40
course/coursed, 1.6.22; 5.7.2
cousin, 1.2.24
cracks, 1.2.37
Crane, Ralph, p. 64
craving, 3.1.34
crests, 5.7.41
cribbed, 3.4.24
Cromwell, Oliver, p. 1
crow, 3.2.54
crowns, 3.4.82
cry of women, 5.5.7.1, 8
Cumberland, Prince of, 1.4.40
Curry, W. C., 2.1.8–10
curtained, 2.1.52
Cusack, Sinead, p. 19
custom, mortal, 4.1.115
cut off, 4.3.79
Cymbeline, pp. 35, 53; 3.1.80

dagger, Macbeth's, p. 4
dainty, 2.3.146
dam, 4.3.218
dance, App. B
Daniel, Samuel, *Delia*, 2.2.34–9
dareful, 5.5.6
darkness in *Macbeth*, pp. 1, 2, 35
Darnley, Henry Stewart, Lord, p. 73
dating of *Macbeth*, pp. 59–66

Davenant, Sir William, pp. 36–40,
 49, 52, 54, 55; 2.1.52; 3.5.33.1;
 3.5.50; 3.5.58; 4.1.43; 4.1.46;
 4.1.52; 4.1.56
de Mille, Cecil B., p. 44
deadman's knell, 4.3.170–1
dear, 5.2.3
death's counterfeit, 2.3.78
dedicate, to greatness, 4.3.75
deed without a name, 4.1.63
deer, 4.3.206
degrees, 3.4.1
delicate, 1.6.10
*demi-wolves, 3.1.94
Dench, Dame Judi, p. 47
Dent, R. W., pp. 86–7; 2.2.59;
 2.3.142–3
devil porters, pp. 79–81
devil's name, th' other, pp. 80–1;
 2.3.7
died, 4.3.111
dignity, 5.1.53
discovery, 5.4.6
dishonours, 4.3.29–30
disjoint, 3.2.17
dismal, 1.2.53
dispatch, 1.5.67; 5.3.49
disposition, 3.4.113–14
dis-seat, 5.3.21
distance, 3.1.116
distempered, 5.2.15
ditch-delivered, 4.1.31
division, 4.3.96
doff, 4.3.188
dollars, 1.2.63
dolour, 4.3.8
Donne, John, *Divine Meditations*,
 4.1.72–4
doom's image, great, 2.3.80
double/doubly, 1.2.38
doubling by actors, pp. 85–6
doubt, 4.2.69
Downes, John, pp. 37–8
downfall birthdom, 4.3.4
drab, 4.1.31
Drayton, Michael, 1.3.11
drenched, 1.7.69
drop of us, each, 5.2.29
Drury Lane Theatre, p. 43
Dryden, John, p. 37; 1.2.38
dudgeon, 2.1.47
Duke of York's Company, pp. 36–7
Duncan, pp. 68–9, 73, 75

dunnest, 1.5.50
Dunsinan Hill, 4.1.108
Dürer, Albrecht, 3.4.101

earnest, 1.3.104
Eckermann, Johann, *Conversations with Goethe*, p. 14
ecstasy, 3.2.24
Edward the Confessor, King, pp. 34, 52, 72, 73; 4.3.140–59
egg, 4.2.85
Eldred, John, pp. 61–2
Eliot, T. S., p. 12
Elizabeth I, Queen, p. 71
Empson, Sir W., 1.3.96; 3.2.49–56; 3.2.54; 3.4.77; 3.6.8; 4.2.22
enkindle, 1.3.122
epicures, English, 5.3.8
equivocates/equivocator, p. 60; 2.3.8; 2.3.33
Erasmus, Desiderius, p. 67; *Colloquia*, 3.1.92–101
Essex, Robert Devereux, 2nd Earl of, pp. 64, 65
estate, 1.4.38; 5.5.50
eternal jewel, 3.1.67
even-handed, 1.7.10
ever, 4.1.117
Evil, the King's, p. 72; 4.3.140–59
Ewbank, Inga-Stina, p. 77
except, 1.2.39
expectation, note of, 3.3.10
expedition, 2.3.112
eye, your, 4.3.186

fact, 3.6.10
faculties, 1.7.17
fair, 1.1.11
family loyalty, p. 75
fantastical, 1.3.53
farmer, 2.3.4
farrow, 4.1.79
fatal, 2.1.37; 2.2.3; 3.5.20–1
father, 2.4.4
fee'd, 3.4.133
fee-grief, 4.3.196
fell (adj.), 1.5.45; 4.3.219; (n.) 5.5.11
fenny, 4.1.12
feverous, 2.3.62
field, 5.1.4
Field, Christopher, p. 38
file, 3.1.95; 5.2.8

filed, 3.1.64
fillet, 4.1.12
Firedrake, 4.1.47
firstlings, 4.1.162
*fit/fits, 3.4.20; 4.2.17
Fitch, Ralph, pp. 61–2
flattering streams, 3.2.36
flaws, 3.4.63
Fletcher, John, 1.3.11
flighty purpose, 4.1.160–1
flout, 1.2.49
Foakes, R. A., 2.2.67–8; 2.3.59; 3.2.52
foisons, 4.3.88
fools, 5.5.22–4
foot of motion, 2.3.126
for to, 1.2.10–12
forbid, 1.3.21
forced, 5.5.5
Forman, Simon, pp. 4, 36, 62, 64, 66; App. C
former tooth, 3.2.16
Fortune, 1.2.14
foul, 1.1.11
founded, 3.4.22
frailties, naked, 2.3.128
frame of things, 3.2.17
franchised, 2.1.29
French, G. R., 5.3.29
French hose, 2.3.13–14
friend, time to, 4.3.10
frieze, 1.6.6
from, 3.1.100
front, 5.7.77
fry, 4.2.86
full, 1.4.55
fume, 1.7.66–8
function, 1.3.141
'function words', p. 57
furbished, 1.2.32

galloglasses, 1.2.13
Garnet, Father Henry, pp. 59–60; 2.3.8
Garrick, David, pp. 41, 42
Gascoigne, George, *Posies*, 2.1.59
genius, 3.1.55
gentle, 1.6.3; 3.4.77
germen, 4.1.72–4
get, 1.3.67
gibbet, 4.1.80
Gielgud, Sir John, p. 47
gild, 2.2.55

gin, 4.2.37
Giulio Romano, p. 6; 5.7.83.1
gives out, 4.3.192
Globe Theatre, pp. 1, 35, 36, 47, 62,
  63, 66, 84; 1.3.78; 3.5.47.1;
  5.7.; App. C
go off, 5.7.66
God 'ield us, 1.6.14
golden round, 1.5.27
Golgotha, 1.2.40
goodness, 4.3.136
goose, roast your, 2.3.14
goose-look, 5.3.12
Gorgon, 2.3.73-4
gouts, 2.1.47
grace/graced, 1.3.55; 3.4.41;
  4.3.23-4; 5.7.102
gracious, 3.1.65
grafted, 4.3.51
Graymalkin, 1.1.8
great doom's image, 2.3.80
green and pale, 1.7.37
green one red, 2.2.61-2
Grieve, T., p. 46
gripe, 3.1.61
grooms, 2.2.5
guise, 5.1.18
gulf, 4.1.23

Hakluyt, Richard, pp. 61-2
half a soul, 3.1.82
Hall, Joseph, *Satires*, 2.3.4-5
Hall, Sir Peter, p. 49
Hamlet, pp. 2, 12, 56, 70, 75;
  1.3.43; 2.3.18; 2.3.36; 3.1.80;
  5.7.61
hand of God, 2.3.132
hangman, 2.2.26
Hanmer, Thomas, p. 52
harbinger, 1.4.46
harness, 5.5.52
harped, 4.1.88
Harpier, 4.1.3
Harris, Anthony, *Night's Black
  Agents*, p. 79
Harrison, William, 2.2.36
haste, 1.2.46
Hattaway, M., p. 63
haunt, 1.6.9
hautboys, p. 36; 1.6.0.1; 1.7.0.1-3
Hawkins, Michael, pp. 75-6
heat-oppressed brain, 2.1.40
heavens, 2.4.5-6

Hecate, pp. 53-5, 57, 64-5; 2.1.53;
  3.2.44
hedge-pig, 4.1.2
Hellway, 3.5.43
hemlock, 4.1.25
Henry IV (1), p. 11
Henry VI (3), 5.7.30
Hercules, p. 58
here approach/remains, 4.3.133;
  4.3.148
hermits, 1.6.21
Hinman, Charlton, p. 50
his, 1.2.14; 1.2.25
hiss the speaker, 4.3.175
hit, 3.6.1
Holdsworth, R. V., pp. 58-9, 63
Holinshed, Raphael, *Chronicle of
  Scotland*, pp. 59, 60-1, 67,
  68-71, 73-4, 75, 77-8; 1.2.63;
  1.3.71; 3.1.133; 4.1.78-9;
  4.3.87; 5.2.0.2; 5.2.2; 5.3.8;
  5.7.83.1; 5.7.93-4; 5.7.100-1
Holland, Philemon, 3.4.102
holp, 1.6.24
home, 1.3.121
honour, country's, 3.4.40
Hopper, 3.5.43
horrid, 1.3.136
horse, 4.1.155
hose, French, 2.3.13-14
Hosley, R., p. 67
hour, slipped the, 2.3.45
house-keeper, 3.1.97
Howard, Frances, Countess of Essex,
  and of Southampton, pp. 36, 64,
  65
howlet, 4.1.17
Hudson, H. N., 2.3.60
human, 3.4.77
Hunter, G. K., 1.2.26; 3.2.52; 5.3.8
hurly-burly, 1.1.3
husbandry, 2.1.5
Hyrcan tiger, 3.4.102

'ield us, God, 1.6.14
Ignatius of Loyola, Saint, *Spiritual
  Exercises*, pp. 26, 28, 32
ignorant, 1.5.56
illness, 1.5.19
illusion in *Macbeth*, pp. 1-6, 23, 32,
  35, 49
imperfect, 1.3.70
imperial, 4.3.20

impress, 4.1.110; 5.7.40
Inchbald, Mrs Elizabeth, p. 43–4
indeed, 3.4.145
informs/informed, 1.5.32; 2.1.49
ingredience, p. 8; 1.7.11; 4.1.34
inhabit, 3.4.106
initiate fear, 3.4.144
insane root, 1.3.84
instruments, 4.3.239
intent, 1.7.26
interdiction, 4.3.107
interim having weighed, the,
  1.3.155
intermission, 4.3.232
intrenchant, 5.7.39
invested, 2.4.32
Irving, Sir Henry, pp. 43, 47
i'th' midst, 3.4.10

Jacobean drama, pp. 31–2
James I, King, pp. 34, 59, 61, 68,
  71–6; 5.7.93; and tyranny,
  pp. 75–6; and witch-hunts,
  pp. 20, 64, 78–9; *Basilikon Doron*,
  p. 75; *Demonology*, pp. 6, 20, 78,
  79; 1.3.124–7
jocund, 3.2.43
Johnson, Robert, p. 65, App. B
Johnson, Samuel, pp. 40–1, 47;
  1.5.52; 2.2.61–2; 3.1.94;
  3.1.130; 3.5.24; 4.2.22; 5.4.11
Jones, Inigo, pp. 24, 34
Jonson, Ben, pp. 10–11; *Eastward
  Ho!*, p. 61; *Masque of Queens*, p. 35
Julius Caesar, pp. 20, 72, 75; 5.7.31
jump, 1.7.7
just censures, 5.4.14
jutty, 1.6.6

Kean, Charles, pp. 44, 45
Kean, Edmund, pp. 44, 47
keen, 1.5.51
Kemble, John Philip, pp. 22, 41,
  43–4
Kemp, William, *Nine Days' Wonder*,
  p. 59
kerns, 1.2.13; 5.7.18
kind/kindness, 1.3.12; 1.5.16
kind'st leisure, 2.1.25
King Arthur (Dryden and Purcell),
  p. 37
King Lear, pp. 20, 32, 53, 56, 59,
  85; 1.3.6; 4.1.72–4; 5.1.6

King's Men, pp. 1, 35, 37, 64, 65,
  71
King's Evil, 4.3.140–59
kites, 3.4.73
Knight, G. Wilson, pp. 12, 23
Knights, L. C., p. 23
knowledge/knowings, 1.2.6; 2.4.4
Kyd, Thomas, *The Spanish Tragedy*,
  4.3.214–15

laced, 2.3.114
Lady Macbeth: acting of, pp. 43, 86;
  children of, p. 14; commentary
  discussion, 1.5.37–53; in
  Holinshed's *Chronicle*, p. 69;
  invocation of spirits, pp. 14–15,
  16, 32; relationship with
  Macbeth, pp. 17–19; sleep-
  walking of, p. 5; soliloquy,
  pp. 14–16; sources for, pp. 77–8
Lady Macduff, pp. 19, 78
Lake, D. J., p. 63 n.
lamp, travelling, 2.4.7
Lancashire, Anne, p. 65
language in *Macbeth*, pp. 7–22
lapped, 1.2.54
large, 3.4.11
largess, 2.1.15
latch, 4.3.195
lated, 3.3.6
lave, 3.2.35
lavish, 1.2.57
lease of nature, 4.1.114
leave/leaves, 4.3.155; 4.3.237
lees, 2.3.97
Leigh, Vivien, p. 48
leisure, kind'st, 2.1.25
Lennox, p. 86; 5.4.0.2–4
Leslie, John, *De Origine Scotorum*,
  pp. 67, 68
Leveridge, Richard, p. 38
levy, 3.2.28
Liard, 4.1.48
lie, giving him the, 2.3.33
life, near'st of, 3.1.118
lighted, 2.3.144
lily-livered, 5.3.15
limbeck, 1.7.66–8
lime, 4.2.37
limited service, 2.3.51–2
line, 1.3.113
lineation in *Macbeth*, pp. 50–1, 86;
  App. A

linen cheeks, 5.3.16
Lingua, p. 63
list, 3.1.70
lizard, 4.1.52
Locke, Matthew, p. 38
lodged, 4.1.69
Loomis, E. A., p. 62
loon, 5.3.11
Louis XIV, King, p. 32
lowliness, 4.3.93
Lucan, *Pharsalia*, 3.5.24
Lucifer, p. 80
Ludus Coventriae, pp. 79, 80
luxurious, 4.3.58
Lyle, E. B., 4.1.140

Macbeth: acting of, pp. 10, 41, 44,
  47; and dagger, p. 4; and
  Macduff's flight, pp. 51, 52; in
  Davenant's version, pp. 38–9; in
  Holinshed's *Chronicle*, pp. 68,
  69–71; relationship with Lady
  Macbeth, pp. 17–19; soliloquy
  (1.7), pp. 7–10
Macduff, pp. 6, 70, 75; in
  Davenant's version, p. 39
McKellen, Ian, pp. 47, 48
Macklin, Charles, p. 43
made a shift to cast him, 2.3.38–9
maggot-pies, 3.4.126
Malcolm, pp. 6, 70, 71, 74, 75
males, 1.7.75
malice, 3.2.15
Malkin, 3.5.57
Malone, Edmond, pp. 59, 72;
  2.3.4–5
man, 1.3.141; 1.7.46; 3.4.58;
  5.7.48
man of blood, 3.4.127
manly, 2.3.135; 4.3.235
*mansionry, 1.6.5
Marlowe, Christopher, p. 85; *Dr
  Faustus*, p. 80; *Edward II*, p. 80
marriage, portrayal in *Macbeth*, p. 19
marshal, 2.1.43
Marston, John, *Eastward Ho!*, p. 61
*martlet, 1.6.4; 1.6.9
marvell'st, p. 18
Mary, Queen of Scots, pp. 59, 73, 74
masques, p. 35
mated, 5.1.76
maw, 4.1.23
Medea, p. 77

memory/memorize, 1.2.40;
  1.7.66–8
mere, 4.3.89; 4.3.152
Merry Wives of Windsor, The, 1.3.6;
  1.3.46
metaphor, p. 10
metaphysical, 1.5.28
Middleton, Thomas, pp. 57–9, 63;
  Black Book, The, p. 63; *Changeling,
  The*, 5.7.64.1–2; *Chaste Maid in
  Cheapside, A*, p. 59; *Game at Chess,
  A*, p. 58; *More Dissemblers besides
  Women*, p. 59; *Puritan, The*, p. 63;
  Widow, The, p. 59; *Witch, The*,
  pp. 54–5, 57, 64–6; 3.5.36;
  3.5.50; 4.1.43; 4.1.56
Middleton, Murry, John, pp. 13–14
midst, i'th', 3.4.10
Midsummer Night's Dream, A, pp. 79,
  85; 5.5.22–4
Miller, Arthur, *The Crucible*, p. 56
Milton, John, *Paradise Lost*,
  1.5.64–5; 2.1.8–10
minion/minions, 1.2.19; 2.4.15
minutely, 5.2.18
miracle plays, pp. 79–81
mischief, nature's, 1.5.49
missives, 1.5.6
Moberly, C. E., 3.2.52
mock the time, 1.7.82
modern, 4.3.170
modest wisdom, 4.3.119
monsters, 5.7.55–6
monuments, 3.4.72–3
more and less, 5.4.12
more-having, 4.3.81
mortal, 1.5.40; 4.1.115; 4.3.3;
  5.3.5
mortified, 5.2.5
motion, foot of, 2.3.126
motives, 4.3.27
mouth-honour, 5.3.27
Much Ado About Nothing, 3.1.80
Muir, Kenneth, pp. 13–14, 15, 51,
  52–3, 63, 67, 76, 84; 1.3.11;
  1.3.141; 1.5.64–5; 1.6.0.1;
  2.1.8–10; 2.1.23–9; 2.1.52;
  2.1.55; 2.2.36; 2.3.72;
  3.1.75–6; 3.4.133–4; 4.1.72–4;
  4.3.214–15; 5.2.2; 5.4.11–12;
  5.5.7.1, 8; 5.7; 5.7.56
mummy, 4.1.23
muse, 3.5.50

music in *Macbeth*, pp. 36, 38;
    3.5.33; App. B
must, 1.6.9

naked frailties, 2.3.128
name, 1.2.16; 2.3.66; 4.1.63
napkins, 2.3.5
Nashe, Thomas, *The Unfortunate
    Traveller*, 2.3.36
National Theatre, p. 49
nature/nature's, 1.5.49; 4.1.114
nave, 1.2.22
navigation, 4.1.68
near/nearest/near'st, 1.4.37;
    1.5.17; 2.3.142; 3.1.118
Newberry, John, pp. 61–2
new-hatched to th' woeful time,
    2.3.60
News from Scotland (1597), p. 78;
    4.1.6–8
Newton, Isaac, p. 31
nice, 4.3.174
nightgown, 2.2.69; 5.1.5
nimbly, 1.6.2
'noint, 3.5.46
noise, 4.1.121
none, 1.7.47
nonpareil, 3.4.18
Norway/Norwegian, 1.2.31, 51
note, 3.3.10; 3.4.56; 5.7.22
nothing is/But what is not,
    1.3.142–3
notion, 3.1.82
Nunn, Trevor, pp. 47, 48

obscure bird, 2.3.60
occasion call us, 2.2.69
offices, 2.1.15
off'rings, 2.1.53
old turning the key, have, 2.3.2
Olivier, Laurence (Lord), pp. 47, 48
on/on's, 1.3.84; 5.1.61
opened, 4.3.52
opera, *Macbeth* as, pp. 34–40, 43–4
opposed, 5.7.61
or ere, 4.3.173
order, 5.6.6
Othello, pp. 12, 70; 1.2.6; 4.3.63;
    5.1.24
other, th', 1.7.28
Other Place, The, Stratford-upon-
    Avon, pp. 47, 48
out, 4.3.183

Overbury, Sir Thomas, p. 64
overcharged, 1.2.37
overcome, 3.4.112
Ovid, p. 79
owe/owed, 1.4.10; 3.4.114
owl, 2.2.3; 2.4.13
Oxford edition (Wells and Taylor),
    pp. 84, 86; 1.2.46; 2.3.92.1;
    2.3.113; 3.5.33.1; 4.1.113;
    5.1.24; 5.7.11.1

paddock, 1.1.9
painting of your fear, 3.4.61
pale, 1.7.37; 3.2.53
pall, 1.5.50
palpable, 2.1.41
palter, 5.7.50
passed in probation, 3.1.79
passion, 3.4.57
patch, 5.3.15
Paul, H. N., *The Royal Play of
    Macbeth*, pp. 72–3, 76
pauser, 2.3.113
pay, 1.3.103; 4.1.114–15
peace, well at, 4.3.179
peak, 1.3.23
pearl, kingdom's, 5.7.86
pendant, 1.6.8
penthouse lid, 1.3.20
Pepys, Samuel, pp. 37, 40
perfect, 4.2.68
pestered, 5.2.23
petty pace, 5.5.20
Phelps, Samuel, pp. 46, 47
physics, 2.3.49
pine, 1.3.23
pious, 3.6.12; 3.6.27
play the woman, 4.3.230
player, 5.5.22–4
plead, p. 9; 1.7.19
plenty, in a spacious, 4.3.71
Pliny, *Natural History*, 3.4.102
plucks, 4.3.119
Poel, William, pp. 46–7
point, 1.2.56; 4.3.135
Polanski, Roman, p. 19
political context of *Macbeth*, pp. 20,
    71–6
poor, 4.2.36, 38; 5.5.24
Pope, Alexander, p. 40; 1.2.26
portable, 4.3.89
Porter, pp. 19, 20, 22, 60; 2.3.1:
    sources for, pp. 79–81

possess them with, 4.3.202
possets, 2.2.6
posters, 1.3.33
power, 2.1.8; 4.3.185; 4.3.236;
  5.6.7
powers above, 4.3.238
practice, 5.1.56
Prayer Book, 5.5.23–4;
  5.5.26–7
prediction, 1.3.55
predominance, 2.4.8
pretence/pretend, 2.3.133; 2.4.24
prick/pricking, 4.1.59; 5.3.14
pride of place, 2.4.12
primrose way, 2.3.18
Pritchard, Mrs, pp. 41, 42
probation, passed in, 3.1.79
profound, 3.5.24
proof, 1.2.54
proper stuff, 3.4.60
proportion, 1.4.19
prospect, 1.3.74
prosperous, 1.3.73; 3.1.21
protest, 5.2.11
proverbial allusions, pp. 86–7;
  1.2.14–15; 1.3.107; 1.3.118;
  1.3.147–8; 1.4.11–12;
  1.4.22–3; 1.5.64–5; 1.6.24;
  1.7.1–2; 1.7.24–5; 1.7.44–5;
  1.7.68; 2.1.6; 2.1.59; 2.2.53–4;
  2.2.59–60; 2.3.30; 2.3.49;
  3.1.80; 3.1.81–2; 3.1.114;
  3.2.12–13; 3.2.58; 3.4.20;
  3.4.22; 3.4.23; 3.4.33–7;
  3.4.67; 3.4.75; 3.4.80; 3.4.123;
  3.4.137–9; 3.5.32–3; 4.1.114;
  4.1.164; 4.2.24; 4.2.42–4;
  4.2.85–6; 4.3.12; 4.3.16;
  4.3.52–3; 4.3.81–2; 4.3.179;
  4.3.204–5; 4.3.216; 4.3.230;
  5.1.65; 5.1.77; 5.3.15; 5.3.16;
  5.5.6; 5.5.23–4; 5.7.1; 5.7.58;
  5.7.69; 5.7.78
Pryce, Jonathan, p. 19
Puckle, 3.5.42
Puckey, 4.1.47
pull in, 5.5.42
punctuation in *Macbeth*, p. 83
Purcell, Henry, pp. 37, 38
Purchas, Samuel, pp. 61, 62
purge, 5.2.28
purveyor, 1.6.23
push, 5.3.20

quarry, 1.2.14; 4.3.206
quarters, 1.3.16
quell, 1.7.73
quenched, 2.2.2
question, 1.3.43

Radcliffe, Mrs., p. 41
rancours, 3.1.66
Rape of Lucrece, The, 2.1.56
rapt, 1.3.57
rather, the, 1.7.63
*ravelled, 2.2.36
raven, 1.5.37
raven up/ravined, 2.4.28; 4.1.24
readiness, manly, 2.3.135
reads, 1.3.90
realism, illusion of, p. 6
reason, 1.7.6–7
rebellious dead, 4.1.112
receipt, 1.7.67
received, 1.7.75
reckon with, 5.7.91
recoil, 4.3.19
Record, of the ... confessions of all the
  Witches, 4.1.46–8
recorded time, 5.5.21
red-haired wench, 4.1.56
reeking, 1.2.39
reflection, 1.2.25
registered, 1.3.152–3
relations, understood, 3.4.125
remember the porter, 2.3.19
remembrance/remembrancer,
  2.3.63; 3.4.37; 5.1.32
rendered, 5.7.25
require, 3.4.6
resolution, 5.5.42
rest, 1.6.21
rhubarb, 5.3.54
Richard II, pp. 11, 72, 75;
  2.3.69–71
Richard III, pp. 10, 12, 56, 76;
  1.3.107
Richardson, Sir Ralph, p. 47
Robin, 4.1.48
Roman fool, 5.7.31
Romeo and Juliet, 2.3.22
*ronyon, 1.3.6
rooks/rooky, 3.2.54; 3.4.126
Ross, pp. 21, 86; 5.4.0.2–3
round, golden, 1.5.27
Rowe, Nicholas, p. 40

Royal Shakespeare Company (RSC), pp. 19, 47–8
Rubens, Peter, *Apotheosis of James I*, pp. 33, 34
rubs, 3.1.134
rump-fed, 1.3.6

sacrilegious murder, 2.3.69–71
Sadler's Wells Theatre, p. 46
safe/safer/safeties, 1.4.27; 3.2.7–8; 4.3.29–30
Saint Colum's Inch, 1.2.62
Saint Matthew, Martyrdom of, pp. 26–7
Sampson, Agnes, pp. 78–9
Sarum Primer, 1.7.11
Satan, pp. 80, 85
satisfy my remembrance, 5.1.32
saucy, 3.4.25
say sooth, 1.2.36
Scaliger, p. 63
scanned, 3.4.141
scarf up, 3.2.50
Scharf, George, p. 45
school, 4.2.15
Scone, 2.4.31
scorched, 3.2.14
Scot, Reginald, *The Discovery of Witchcraft*, pp. 20, 66, 78, 79; 3.4.72–3
scrofula, p. 72; 4.3.140–59
scruples, 2.3.131
season, 3.4.142; 4.2.17
seat/seated, 1.3.137; 1.6.1
second cock, 2.3.22
second course, 2.2.38
secret'st man of blood, 3.4.127
security, 3.5.32
seeds, 3.1.69
seeling, 3.2.49
seems to, 1.2.47
self-abuse, 3.4.143
self-comparisons, 1.2.55
Seneca, pp. 66, 76–7; *Agamemnon*, p. 77; *Hippolytus*, 4.3.209–10; *Medea*, pp. 77, 79
senna, 5.3.54
sennet, 3.1.10.1
sense, 5.1.24
sensible, 2.1.37
sere, 5.3.23
sergeant, 1.2.3
service, 1.7.0.3; 2.3.51–2

setting down, 5.4.10
sewer, 1.7.0.2
sexuality in *Macbeth*, p. 19
Seyton, p. 85; 5.3.29
Seyward, 4.3.134; 5.2.2
shadow, 5.5.22–4
shag-eared, 4.2.85
shard-born, 3.2.45
shift away, 2.3.147
Shirley, Frances, *Shakespeare's Use of Off-stage Sounds*, pp. 35–6
shoal, 1.7.6
shook hands, 1.2.21
*shoughs, 3.1.94
shut up, 2.1.17
Siddons, Sarah, pp. 15, 42, 43–4; 1.5.35–51; 1.5.60; 1.7.60–1
Sidney, Sir Philip, *Astrophel and Stella*, 2.2.34–9; 2.2.37
sieve, 1.3.8
sightless, pp. 9, 16; 1.5.48; 1.7.23
Simon, Irène, 1.3.6
Sinell, 1.3.71
single, 1.3.141; 1.6.17
skipping, 1.2.30
skirr, 5.3.34
slab, 4.1.32
sleep-walking, p. 5
sleeve, 2.2.36
slipped the hour, 2.3.45
slips of yew, 4.1.27
slivered, 4.1.28
slope, 4.1.71
smoked, 1.2.18
society, 3.4.4
sole/solely, 1.5.69; 4.3.12
solemn, 3.1.14
soliciting, 1.3.131
something, 3.1.132
Sonnets, 5.1.24; 5.7.48
songs, App. B
sooth, 1.2.36; 5.5.40
sore, 2.4.3
sorts, 1.7.33
sound effects, pp. 35–6
sources for *Macbeth*, pp. 66–81
Southampton, Henry Wriothesley, 3rd Earl of, p. 64
sovereign, 5.2.30
sows, 4.1.78–9
spacious plenty, in a, 4.3.71
speculation, 3.4.96
speed of, had the, 1.5.34

spelling in *Macbeth*, p. 83
Spenser, Edmund, *The Faerie Queene*,
    3.4.72–3
spirits, pp. 14–16; 1.5.25;
    1.5.39–40; 5.3.4
spoken, 4.3.154
spongy, 1.7.72
spring, 1.2.27
sprites, 2.3.81
spy o' th' time, perfect, 3.1.130
Stadling, 3.5.41
staff, 5.3.47
staging of *Macbeth*, pp. 34–49
stamp, 4.3.153
stanchless, 4.3.78
stand in thy posterity, 3.1.4
stand to, 2.3.32
state, 3.1.33, 3.4.5
state of honour, 4.2.68
station, 5.7.72
Stationers' Register, The, p. 59
staves, 5.7.18–19
steals, 2.3.148
steel, 1.2.17
steeped, 2.3.116–8
Steevens, George, 1.3.9; 3.1.74;
    3.2.52; 3.5.24; 5.3.21
stern'st good night, 2.2.4
Stewart, William, *The Buik of the
    Chronicles of Scotland*, p. 68
sticking, 1.7.61; 5.2.17
still, 3.1.21
stones prate, 2.1.59
stool/stools, 3.4.68; 3.4.83
strangles, 2.4.7
striding, 1.7.22
strike beside us, 5.7.30
Stuart genealogy, pp. 68, 73–4
studied, 1.4.9
stuff/stuffed, 3.4.60; 5.3.43
suborned, 2.4.24
substances, 1.5.48
sudden, 4.3.59
summer-seeming, 4.3.86
summons, 2.1.7
supernatural in *Macbeth*, pp. 3, 6,
    23–4
surcease, 1.7.4
surgery, 4.3.152
surmise, 1.3.142
surveying vantage, 1.2.31
sway, 5.3.9
sweaten, 4.1.79

swelling, 1.3.129
sweltered, 4.1.8
swinish sleep, 1.7.68
swoop, fell, 4.3.219
syllable, 4.3.8; 5.5.21

tailor, English, 2.3.13–14
taint, 5.3.3
takes ... off, 3.1.105
Tarquin, 2.1.56
Taylor, Gary, pp. 57, 83; 4.3.59
tears, p. 9
teems, 4.3.176
Tempest, The, pp. 5–6, 35, 53;
    5.7.55–6
tending, 1.5.36
Teresa of Avila, Saint, *Life*,
    pp. 26–31
Terry, Dame Ellen, p. 43
text of *Macbeth*, pp. 49–56,
    App. A
Thane, 1.2.45; 1.3.120
theirs, 1.6.27
theme, 1.3.130
Theobald, Lewis, p. 40
thick my blood, make, 1.5.42
thickens, 3.2.53
thirst, 3.4.92
Thomas, Keith, *Religion and the
    Decline of Magic*, pp. 67, 78;
    1.2.53; 1.3.9; 1.3.11; 1.3.46
thought, upon a, 3.4.55
thrice, 1.3.35
thriftless, 2.4.28
throat on me, 2.3.36
Tiffin, 4.1.46
Tiger, pp. 61–2; 1.3.7
Tiger's Whelp, p. 62
Tilley, M. P., *Dictionary*, pp. 86–7;
    2.2.59–60; 2.3.142–3
time, 1.5.62; 1.7.82; 2.3.60;
    3.1.130; 4.1.114–15; 4.3.10;
    4.3.235; 5.5.21; 5.7.85
timely, 2.3.44
Timon of Athens, p. 58
Titivillus, p. 80
title/titles, 4.2.7; 4.3.34
Titty, 4.1.46
Titus Andronicus, p. 76
toad, 4.1.6–8
took up my legs, 2.3.38
tooth, former, 3.2.16
torch, 1.6.0.1; 1.7.0.1; 2.1.0

Tourneur, Cyril, *The Atheist's Tragedy*, pp. 64, 65
Towneley miracle plays, p. 80
tow'ring, 2.4.12
trace, 4.1.168
trains, 4.3.118
trammel up, 1.7.3
transpose, 4.3.21
travelling lamp, 2.4.7
treasure, 4.1.72–4
treatise, 5.5.12
Tree, Sir Herbert Beerbohm, p. 47
trenched, 3.4.27
trifled former knowings, 2.4.4
Troilus and Cressida, 3.2.17
trouble, 1.6.11–15
trumpet, p. 36; 2.3.83–4
tugged with, 3.1.112
turrets, 3.5.63
tyranny, pp. 20, 21–2, 75–6; 4.3.67

unattended, 2.2.67–8
unbend, 2.2.44
undeeded, 5.7.21
understood relations, 3.4.125
undivulged pretence, 2.3.133
unfix, 1.3.136
unrough, 5.2.10
unseamed, 1.2.22
unsex, 1.5.40
untimely ripped, 5.7.46
untitled, 4.3.104
upon his aid, 3.6.30
uproar, 4.3.99
use/using, 1.3.138; 3.2.11
utterance, 3.1.71

valued file, 3.1.95
vantage, 1.2.31; 1.3.114; 1.6.7
vap'rous drop, 3.5.24
vault, 2.3.98
vaulting, 1.7.27
venture, 1.3.91
verbal illusion, p. 6
Verdi, Guiseppe, pp. 44, 46
vessel/vessels, 3.1.66; 3.5.18
viols, p. 36
Virgil, *Eclogues*, 1.5.64–5
virtue, 4.3.156
visited, 4.3.150
vizards, 3.2.37

Walpole, Horace, p. 41
want, 3.6.8
wanton, 1.4.35
Warburton, William, p. 60; 2.1.57–61
Warner, William, *Albion's England*, p. 71
warrant/warranted, 2.3.147; 4.3.137
wassail, 1.7.65
wasteful, 2.3.116
watch/watchers/watching, 2.1.55; 2.2.70; 5.1.10
water-rugs, 3.1.94
way and move, 4.2.22
way of life, 5.3.22
wayward, 3.5.11
we, 1.5.23–9; 5.7.91
weal, gentle, 3.4.77
Webster, John, *The Duchess of Malfi*, 2.2.3
Weekes, John, *Truth's Conflict with Error*, 1.7.11
weird, p. 3
Weïrd Sisters: and Hecate, pp. 53–4; and illusion, pp. 2–4, 5, 6; and witch-hunts, pp. 20, 78–9; commentary discussion, 1.1.0; 1.3.32; 1.3.39–47; language of, p. 22; sources for, pp. 61, 69, 76, 78–9
well at peace, 4.3.179
Wells, Stanley, *Modernizing Shakespeare's Spelling*, p. 83
Western Isles, 1.2.12
what is the night?, 3.4.127
while, 3.1.44
white, 2.2.64
will, 4.3.65
Wilson, J. Dover, pp. 14, 57; 1.6.0.1
Wilson, John, p. 65, App. B
wind, 1.3.11; 1.7.24–5
wink, 1.4.53
Winter's Tale, The, pp. 1, 6, 35; 1.5.56; 5.3.83.1
wit, 4.2.45–6
Witch/Witches, 1.1.0; 1.3.6; *see also* Weïrd Sisters
witch-trials, pp. 20, 64, 78–9
withal, 2.1.16
within, 2.3.0; 3.5.35.1
without/within, 3.4.14; 4.1.150
Woodvine, John, p. 48

worlds, both the, 3.2.18
worst means, 3.4.136
wound up, 1.3.37
wrack, 1.3.115; 5.5.51
written, 5.3.41

wrought, 1.3.150; 1.5.20; 2.1.20;
3.1.81
yeasty, 4.1.67
York miracle plays, pp. 79–80
younker, 4.1.54

Index compiled by Peva Keane